A. Lee Carter

CURRENT CONCEPTS

IN

CARNITINE RESEARCH

CURRENT CONCEPTS IN CARNITINE RESEARCH

Editor

A. Lee Carter
Associate Professor
Department of Biochemistry
and
Molecular Biology
Medical College of Georgia
Augusta, Georgia

CRC Press
Boca Raton   Ann Arbor   London

Library of Congress Cataloging-in-Publication Data

Catalog record is available from the Library of Congress.

International Standard Book Number 0-8493-0186-6

Printed in the United States   1 2 3 4 5 6 7 8 9 0

TABLE OF CONTENTS

Preface

This book contains the proceedings of the latest in a series of symposiums on basic research with carnitine. The first symposium was in 1965. As with the other symposium books, there is much new data presented. Since the areas of carnitine research are too numerous to be covered in detail, three areas were chosen for detailed investigation. One section contains investigations into functions, synthesis, and metabolism of acylcarnitines. Another section contains several viewpoints into the localization and function of the carnitine palmitoyl transferases. A third section has as it's emphasis the role of carnitine in intermediary metabolism of cardiac muscle. A fourth section is an introduction to new areas of carnitine research. This section is intended to stimulate research into these new areas of carnitine research.

Therefore, this book is a good reference for the areas of carnitine research which are covered in detail and an introduction into new areas of carnitine research which may become important in the near future.

The Editor

Dr. A. Lee Carter was born in Uvalde, Texas on June 2, 1948. After his childhood years in Camp Wood, Texas, he received a bachelor of Science degree in Chemistry from Baylor University in 1971. A doctor of philosophy degree in biochemistry was granted in 1977 from the University of Texas Health Science Center at Dallas. Under the director of Dr. Rene A. Frankel, his dissertation was titled "Studies on the Biosynthesis and Transport of Carnitine and its Precursors in the Rat".

After a three year post-doctoral fellowship under Dr. Henry Lardy at the Enzyme Institute, University of Wisconsin at Madison, Madison Wisconsin; he accepted a position as Assistant Professor in the department of Cell and Molecular Biology at the Medical College of Georgia. He is currently Associate Professor in the department of Biochemistry and Molecular Biology at the Medical College of Georgia.

Dr. Carter's research career has been with the study of carnitine metabolism and he is particularly interested in the determination of carnitine levels and factors which affect both the biosynthesis of carnitine and the regulation of tissue and whole body carnitine levels.

ACKNOWLEDGEMENTS

This symposium was made possible by an educational grant from Sigma Tau Pharmaceutical Company to the Medical College of Georgia. The success of this symposium, and subsequently this book, depended on the hard work and dedication of a great number of people. For the planning and execution of the symposium itself, I would like to thank the folks from the office of Continuing Education - Dr. Gerald Chambers, Ms. Katrinka J. Akeson, Ms. Shannon E. Shelton, and Ms. Laura E. Hennessey. The monumental task of helping me put these proceedings into publishable form was undertaken superbly by Ms. Elisabeth R. Bishop, Ms. Ann P. Patch, and Ms. Marcia DeLandro Lewis. Elwood N. Longenecker compiled the index for this volume. Last, but not least, I wish to thank the participants in this symposium. Without you, none of this would ever have happened!

PART I

ACYL CARNITINE METABOLISM

INTRODUCTION

James E. Carroll, M.D.
Depart of Neurology (Pediatric Neurology)
Medical College of Georgia
Augusta, GA 30912

It is appropriate that we begin this conference with a discussion of acylcarnitine metabolism. The term "acylcarnitine" highlights the importance of carnitine as a transporting agent for various species of acyl groups across membranes, most notably those in mitochondria and renal tubules. These two basic mechanisms of carnitine function are critical both in regard to the so-called detoxifying role of carnitine in mitochondria and also to the loss of carnitine that occurs from the body across the renal tubule as acylcarnitine. We will briefly examine these two issues as examples of current problems in carnitine research.

Detoxifying Role of Carnitine

High concentrations of abnormal acyl-CoA derivatives may accumulate in conditions affecting organic acid and fatty acid metabolism, including propionic acidemia, isovaleric acidemia, methylmalonic acidemia, glutaric acidemia, 3-ketothiolase deficiency, 3-hydroxy-3-methylglutaryl-CoA lyase deficiency, medium chain acyl-CoA dehydrogenase deficiency, multiple acyl-CoA dehydrogenase deficiency, multiple carboxylase deficiency, and valproate therapy[6]. Their respective acyl-CoA derivatives are thought to produce adverse effects at the mitochondrial level by disruption of the normal acyl-CoA and CoASH concentrations and by inhibition of certain enzymatic reactions. The administration of carnitine is proposed to reverse these two processes by the following exchange:

$$\text{Acyl-CoA} + \text{carnitine} \rightleftarrows \text{acylcarnitine} + \text{CoASH}$$

The resulting acylcarnitine is then eligible for transport across the mitochondrial membrane via a translocase and eventual excretion.

There have been few measurements of tissue CoASH in these diseases before or after carnitine therapy. In experimental circumstances where this determination has been carried out, the concentration of CoASH in the disease state is still higher than that employed in most enzyme assays in vitro[8]. Thus, the simple depletion of CoASH is insufficient to account for the adverse effects of the accumulating acyl-CoA derivatives; the acyl-CoA/CoASH ratio may be of greater importance.

The more serious effects of these acyl-CoA derivatives are likely to be mediated through their inhibition of enzymes such as alpha-ketoglutarate dehydrogenase, pyruvate dehydrogenase, citrate synthase, pyruvate carboxylase, carbamylphosphate synthetase, and N-acetylglutamate synthetase[2]. Clinically, we see resulting hypoglycemia, lactic acidemia and hyperammonemia.

Carnitine is recommended for its role in correcting both the diminished CoASH concentration and the elevated concentrations of abnormal acyl-CoA metabolites. In a patient with isovaleric acidemia treated with carnitine, however, urinary excretion of isovalerylcarnitine accounted for less than 5% of the calculated isovaleryl-CoA formed[3]. This result suggests that carnitine is not an efficient acceptor of toxic acyl groups and that its clinical efficacy may be minimal. Similarly, in medium chain acyl-CoA dehydrogenase deficiency, carnitine supplementation failed to prevent the toxic effects of accumulating medium chain fatty acids during fasting[9].

Considering the relative rarity of the diseases generating toxic acyl metabolites and their genetic heterogeneity, the paucity of controlled studies to investigate the benefit of carnitine therapy in these conditions is not surprising.

Renal Excretion of Acylcarnitine

Most organic aciduria patients demonstrate "secondary" carnitine deficiency. Since the main excretory form of carnitine is acylcarnitine, and since acylcarnitine plasma concentrations are elevated in organic acidurias and fatty oxidation disorders, low tissue carnitine in these disorders is thought to result from excessive urinary loss of carnitine as acylcarnitine.

Carnitine is filtered through the glomerulus and reabsorbed at the level of the proximal tubule. Additionally, since the renal excretion rate of total carnitine can exceed the glomerular filtered load, carnitine is probably secreted into the tubular lumen[4]. In the normal circumstance, most carnitine is excreted in the urine as short chain acylcarnitine, mainly acetylcarnitine. Acetylcarnitine/carnitine ratios in urine exceed those in plasma. Even higher proportions of acylated carnitine derivatives are found in urine in organic acidurias. Mitochondrial and cytosolic carnitine acetyltransferases activities are high in the kidney.

These observations have suggested that acylation of carnitine may take place in the renal tubule. Branched-chain acylcarnitines are formed and excreted in urine from kidneys perfused with carnitine and branched chain amino acids[5]. Isolated kidney tubules incubated with carnitine and substrates such as lactate, acetate, oleate, acetoacetate, and 3-hydroxybutyrate produce acylcarnitine. Thus, Guder and Wagner have concluded that the proximal tubule itself actively produces new carnitine esters[4]. These data suggest that the substrates, carnitine and acyl groups, must be available to the kidney in order for acylcarnitine excretion to occur. This availability occurs more prominently during fasting.

Collection of total urinary carnitine in these patients usually reveals low rather than high excretion[7]. In fact, their urinary excretion may be less than their intake; excretion is almost always greater than intake in normal individuals[1]. Only during fasting, when more acyl groups are available in these patients, does the carnitine excretion rise to levels which vastly exceed the normal rate[7]. Since such episodes are uncommon and are usually avoided, it seems unlikely from these data that urinary loss alone could account for the low levels of tissue carnitine found in these patients. Other possibilities, such as decreased synthesis, should be entertained. We should also note, however, that plasma and tissue levels are nearly always low by the time the patients are diagnosed, and therefore the total body pool of carnitine is already decreased. We obviously do not know what transpired in terms of urinary excretion while the patients were establishing their low carnitine status. Clearly, the final answer is not available as to the mechanism for reaching the low tissue carnitine levels.

As carnitine is replaced in these patients, urinary excretion may rise considerably and often tissue levels rise only marginally. Is the rise that occurs in tissue and plasma significant from a biochemical standpoint? That is, does the rise in carnitine enhance the rate of relevant enzymatic reactions? Is there clinical improvement under controlled evaluation?

These two brief examples suffice to show that there are simple, but unresolved issues in carnitine science. I urge that we regularly address two questions: <u>Does the proposed mechanism make good biochemical sense? Is the mechanism clinically significant?</u>

REFERENCES

1. Carroll JE, Brooke MH, Shumate JB, Janes NJ: <u>Am J Clin Nutr</u>, 34: 2693-2698, 1981.
2. Corkey BE, Deeney JT: Fatty Acid Oxication: Clinical, Biochemical, and Molecular Aspects, <u>Liss, Inc.</u>, pp. 217-232, 1990.
3. DeSousa C, Chalmers RA, Stacey BM, Tracey BM, Weaver CM, Bradley D: Eur J Pediatr, 144: 451-456, 1986.
4. Guder WG, Wagner S: J Clin Chem Clin Biochem, 28: 347-350, 1990.
5. Hokland BM, Bremer J: Biochim Biophys Acta, 886: 223-230, 1986.
6. Roe CR, Millington DS, Kahler SG, Kodo N, Norwood DL: Fatty Acid Oxidation: Clinical, Biochemical, and Molecular Aspects, <u>Liss, Inc.</u>, pp. 383-402, 1990.
7. Schmidt-Sommerfield E, Penn D, Kerner J, Bieber LL, Rossi TM, Lebenthal E: J Pediatr, 115: 577-582, 1989.
8. Sherrat HSA: Biochemical Society Transactions, 13: 856-858, 1985.
9. Treem WR, Stanley CA, Goodman SI: J Inher Metab Dis, 12: 112-119, 1989.

ACYLCARNITINE FUNCTION AND ENZYMOLOGY

L. L. Bieber, G. Dai, and C. Chung
Department of Biochemistry
Michigan State University
East Lansing, MI 48824

I. INTRODUCTION

Most mammalian tissues have the capacity to convert 2-carbon to > 20-carbon acyl-CoAs to acylcarnitine. The reactions are catalyzed by a family of enzymes, the carnitine acyltransferases, which reversibly transfer the acyl residue of acyl-CoA to the β-hydroxyl of carnitine. The fact that these enzymes show a broad acyl-CoA specificity with a similar V_{max} for a range of acyl-CoAs, has led to the same enzymes being referred to by different names. Herein, CAT (carnitine acetyltransferase) is the enzyme which has a high V_{max} for short-chain acyl-CoAs as substrate. COT (carnitine octanoyltransferase) is the carnitine acyltransferase that uses both medium-chain and long-chain acyl-CoAs as substrates, but has a higher V_{max} with medium-chain than with long-chain acyl-CoAs, i.e., those enzymes where application of standard biochemical criteria for nomenclature of enzymes, namely comparison of the $K_{0.5}/V_{max}$ for 8- and 10-carbon acyl-CoAs versus 16-carbon acyl-CoAs, demonstrates a kinetic preference for medium-chain acyl-CoAs. These enzymes are primarily associated with microsomes and/or peroxisomes.[1,2] CPT (carnitine palmitoyltransferase) is the mitochondrial enzyme that transfers medium-chain and long-chain acyl-CoAs to carnitine, but it shows a greater V_{max} with long-chain acyl-CoAs at physiological concentrations of carnitine.[3]

It has been evident for many years that carnitine has functions other than β-oxidation of long-chain fatty acids in some tissues. Some of the clearest evidence has come from systems which do not oxidize long-chain fatty acids, but contain large amounts of CAT and/or require L-carnitine for growth. In a similar meeting about 10 years ago, it was proposed that several different functions for carnitine are evident from the literature.[4] With the recognition that during some human disease states, carnitine deficiency can apparently be induced due to a metabolic stress, considerable interest in the non-β-oxidation roles for carnitine arose.[5-7] As a result of this, evidence accumulated which demonstrated that carnitine roles include not only the well-defined one in mitochondrial β-oxidation of fatty acids, but other functions such as detoxification of non-metabolizable acyl residues, general buffering of the acyl-CoA to CoA ratio, and the carnitine-mediated shuttling of the acyl groups of acyl-CoAs between mammalian tissue compartments, such as peroxisomes, cytosol, and mitochondria.

II. METHODS

For the exchange reaction, 50 μM acylcarnitine as substrate and/or inhibitor were used in a media that contained 20 mM P_i,, pH 7.5, 0.4 mM DTT, 0.08 mM CoASH, and 0.25 μC methyl-^{3}H-L-carnitine. Incubations were at 35° for 30 minutes to 3 hours, depending on the experiment done. HPLC assays were essentially as described.[17] Other methods and procedures are described and/or cited in the figure and table legends.

III. RESULTS and DISCUSSION

A. NOT ALL CATS ARE KINETICALLY ALIKE

Although all CATs reversibly transfer short-chain acyl moieties between CoASH and carnitine, their specific function in different tissues varies, and their acyl chain length specificities can vary. For example, the organism which requires carnitine for growth has a CAT that shows very narrow acyl chain length specificity; no catalytic activity with acyl chain lengths greater than 4-carbons was detected.[8,9] In contrast, most mammalian CATs have a broad acyl chain length specificity, with a greater V_{max} for propionyl, and sometimes butyryl, than for acetyl-CoA.[10] CAT exhibits both Michaelis-Menten kinetics and non-Michaelis-Menten kinetics. As shown in Figure 1, rat heart mitochondrial CAT has a Hill coefficient of 1.6 with acetyl-CoA as substrate, but near 1.0 when propionyl-CoA is the substrate. This CAT (data submitted for publication) shows a lag in acetylcarnitine production when acetyl-CoA is substrate. In contrast, when acetyl-CoA is replaced by propionyl-CoA, Michaelis-Menten velocity versus substrate concentration responses are obtained. This observation provides an explanation for previous results which showed a lag in the production of acetylcarnitine by heart mitochondria when the acetyl-CoA to CoA ratio was low.[12,13] These studies indicate that mitochondrial CAT has unique properties which allow conservation of acetyl-CoA for the TCA cycle when mitochondrial matrix acetyl-CoA is limited.

B. FUNCTIONS AND PROPERTIES OF COT

Mitochondria, peroxisomes and microsomes of liver each contain a carnitine acyltransferase that catalyzes the conversion of medium-chain and long-chain acyl-CoAs to acylcarnitine. Considerable differences occur in the literature in the naming of these enzymes. Herein, the microsomal and peroxisomal enzymes are referred to as COTs (carnitine octanoyltransferases) because the $V_{max}/K_{0.5}$ for C10 acyl-CoAs compared to C16 acyl-CoAs for the microsomal enzyme and the peroxisomal enzyme are approximately 30 and 10, respectively, demonstrating that the enzymes have a kinetic preference for medium-chain acyl-CoAs as substrate. Although the V_{max}s are quite different, the $K_{0.5}$s for acyl-CoAs are similar, being in the low micromolar range, and the $K_{0.5}$s for L-carnitine are within the physiological range for carnitine.[1,11,14] Sensitivity to malonyl-CoA has been reported for both the microsomal enzyme and the peroxisomal enzyme;[15,16] thus, it appears that cytosolic conversion of acyl-CoAs to acylcarnitines by microsomal COT does not bypass the short-term metabolite control by malonyl-CoA that regulates the conversion of acyl-CoAs to acylcarnitines.

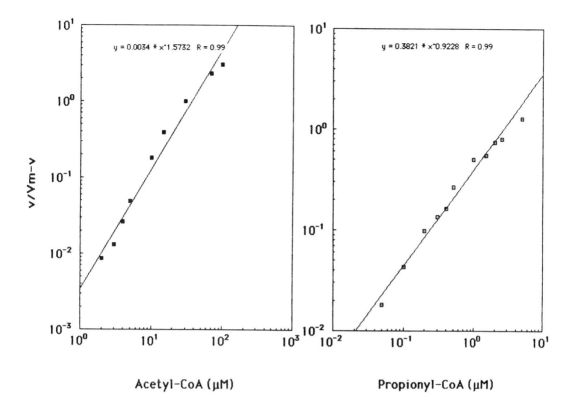

Figure 1. **Hill coefficients for rat liver mitochondrial CAT.** _CAT was purified using standard chromatographic techniques (ammonium sulfate precipitation, Sepharose 6B, and carboxymethylcellulose chromatography) to a specific activity of 11.9 U/mg protein with 37% recovery. Kinetic analyses at pH 7.5 and 3.8 mM L-carnitine were performed as described elsewhere._[3,10,11]

The function for the medium-chain COT in liver microsomes is not known. Considerable evidence has accumulated that shows peroxisomes have the capacity to convert the medium-chain acyl-CoAs, produced by peroxisomal β-oxidation of long-chain acyl-CoAs, to medium-chain acylcarnitines. It is plausible that these medium-chain acylcarnitines in the cytosol are reconverted to medium-chain acyl-CoAs by the microsomal transferase (the green enzyme shown in Figure 2) to subsequently be used as substrates for microsomal enzymes such as chain elongation and α- and ω-oxidation. However, definitive evidence for this potential role has not been obtained.

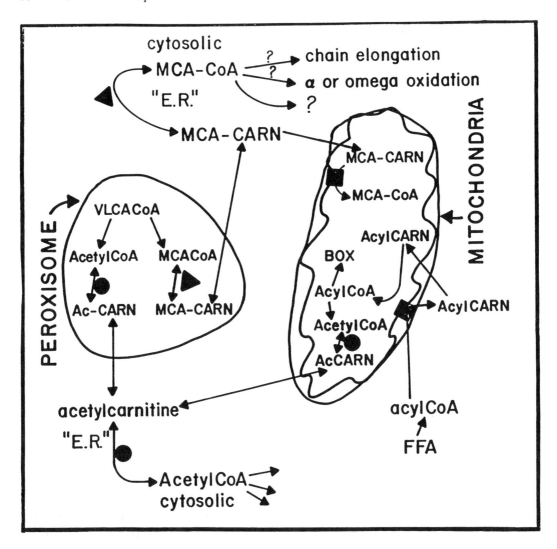

Figure 2. **Potential sites of medium-chain acylcarnitine utilization and production.** *The circles = CAT. The triangles = COT. The squares = CPT. ER = endoplasmic reticulum; MCA = medium-chain acyl; VLCA = very long-chain acyl; CARN = carnitine; FFA = free fatty acids; BOX = β-oxidation.*

C.   ATYPICAL ACYLCARNITINES

The capacity of several carnitine acyltransferases to form short-chain acylcarnitine derivatives, including atypical ones, has been investigated. Both COT and CAT can convert pivaloyl-CoA to pivaloylcarnitine, but they do not show any catalytic activity towards glutaryl- and valproyl-carnitine. As shown in Figure 3, pivaloyl-CoA is

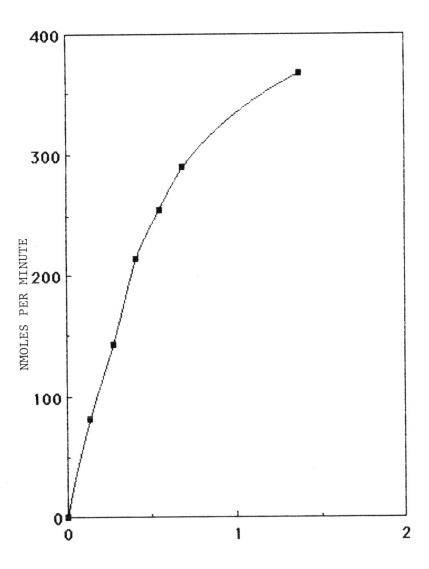

PivaloylCoA (mM)

Figure 3. Velocity versus pivaloyl-CoA concentration profile for rat liver peroxisomal CAT. CAT was assays spectrophotometrically in the forward direction using 1.25 mM L-carnitine at pH 7.5.[1,3]

converted to pivaloylcarnitine with a $K_{0.5}$ = 1.0 mM and a Hill n $\cong$ 0.95. Although pivaloyl-CoA is a substrate for CAT, the K_m is > 300-fold higher than the $K_{0.5}$ for acetyl-CoA; compare 1000 μM to 3 μM. Thus, pivaloyl-CoA is a very poor substrate and must be present in high concentrations relative to acetyl- and other short-chain acyl-CoAs if it is to be converted to acylcarnitine. In contrast, valproyl- and glutaryl-carnitine do not serve as substrates for the reverse reaction, although pivaloyl-, valproyl-, and glutaryl-carnitine inhibit both CAT and COT; thus, they must bind to the enzymes, but do not undergo catalysis. Figure 4 shows the effect of valproylcarnitine and pivaloylcarnitine on the conversion of acetylcarnitine and octanoylcarnitine to acyl-CoAs by peroxisomal CAT and COT, respectively.

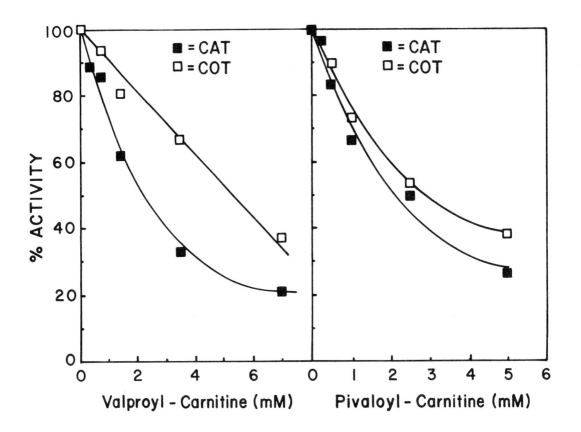

Figure 4.   **Inhibition of peroxisomal CAT and COT by valproylcarnitine and pivaloylcarnitine.**   *Purified CAT and COT were assayed spectrally in the reverse direction at 233 nm with 500 μM acetyl- or octanoyl-carnitine and 400 μM CoASH at pH 7.5.*

Since some of the unusual acylcarnitines can serve both as substrates for and inhibitors of carnitine acyltransferases, the effect of these compounds on individual carnitine acyltransferases was determined using the isotope exchange assay.[17] With this method, conversion of substrate to product, the conversion of the inhibitor to a product and the inhibition of the reaction can be simultaneously determined. This is illustrated in Figure 5, in which Figure 5A shows the exchange of ^{3}H-carnitine into acetyl-L-carnitine by liver mitochondrial CAT; addition of pivaloylcarnitine to the reaction mixture both inhibits exchange of carnitine into acetylcarnitine and results in a small exchange of ^{3}H-carnitine into pivaloylcarnitine. Thus, it is a poor substrate and a modest inhibitor of the enzyme (see Figure 5C). Addition of methylglutarylcarnitine (Figure 5B) results in strong inhibition of the exchange reaction; thus, methylglutarylcarnitine must bind to the active site, but it does not undergo catalysis. Figure 5D shows the effect of

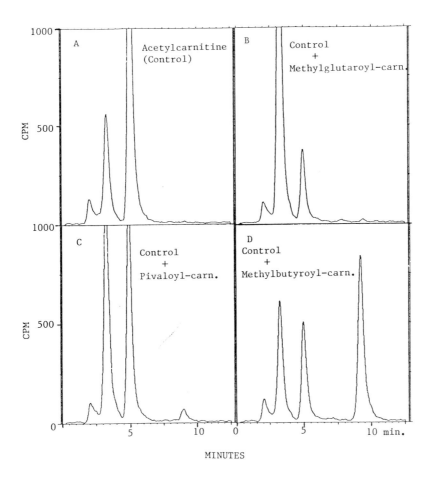

Figure 5. **Effect of valproylcarnitine and pivaloylcarnitine on the exchange of carnitine into acetylcarnitine. Fig. 5A = control. Fig. 5B = control + equimolar pivaloylcarnitine. Fig. 5C = control + equimolar valproylcarnitine. Fig. 5D = control + equimolar 2-methylbutyrylcarnitine.**

2-methylbutyrylcarnitine, which is both a substrate and an inhibitor. The results of these experiments are summarized in Table 1. Interestingly, 3-methylglutarylcarnitine potently inhibits liver CATs when acetylcarnitine is the substrate, but not when 2-methylbutyryl-carnitine is the substrate. Although we have not obtained evidence which shows that any of the rat tissue enzymes use valproyl-, glutaryl- or 3-methylglutaryl-carnitine as substrates, the data clearly show that these compounds can inhibit some of the rat carnitine acyltransferases, especially liver CATs. In contrast, as shown in Table 1, rat heart mitochondrial CAT shows little inhibition by the unusual acylcarnitines. Thus, in organisms such as humans which form methylglutarylcarnitine and valproylcarnitine derivatives, one can infer from the studies with rat tissues that these derivatives should have inhibitory effects on some of the carnitine acyltransferases. It is not known whether such inhibitory effects contribute to any of the metabolic effects encountered in the human disease states where these acylcarnitines are excreted, and it is not known which enzymes produce the unusual urinary acylcarnitines.

TABLE 1
Effect of Acylcarnitines on the Exchange of ^{3}H-L-Carnitine into Acetyl-, Octanyl-, 2-Methylbutyryl-, and Decanyl-Carnitines by Different Carnitine Acyltransferases

Enzyme (Source)	n	Pivaloyl-	Valproyl-	2-Methylbutyryl-	3-Methylglutaryl-	Glutaryl-
				(% of control)		
Acetyl- as Substrate						
CAT (rat liver peroxisome)	3	69.4 ± 3.8	96.4 ± 4.6	40.6 ± 5.7	26.8 ± 2.9	87.1 ± 6.5
CAT (rat liver mitochondria)	3	68.3 ± 2.1	98.0 ± 2.6	39.3 ± 4.4	25.4 ± 3.1	77.8 ± 16.1
CAT (pigeon breast muscle)	3	96.7 ± 3.5	97.1 ± 8.3	61.1 ± 7.5	97.3 ± 5.1	87.0 ± 6.8
CAT (rat heart mitochondria)	3	100.6 ± 13.6	101.8 ± 3.7	82.4 ± 7.6	100.1 ± 3.1	97.3 ± 7.6
Octanyl- as Substrate						
COT (rat liver microsomes)	3	98.6 ± 2.8	101.3 ± 4.0	50.1 ± 14.9	105.9 ± 10.6	107.1 ± 17.3
Decanyl- as Substrate						
CPT (rat heart mitochondria)	3	93.2 ± 8.3	98.9 ± 1.9	86.2 ± 2.6	101.8 ± 2.6	100.0 ± 5.5
2-Methylbutyryl- as Substrate						
CAT (rat liver peroxisome)	4	99.8 ± 16.6	107.1 ± 5.4	---	106.5 ± 5.4	96.2 ± 11.1
CAT (rat liver mitochondria)	4	92.1 ± 13.5	90.1 ± 18.9	---	98.4 ± 9.4	86.1 ± 13.0

• Equal concentrations of substrate and acylcarnitine were used.

D.   FUNCTION OF MITOCHONDRIAL CPT

The role for mitochondrial CPT in the β-oxidation of long-chain fatty acids is well-established and is a topic in this conference; it will not be reviewed here; however, it should be remembered that although CPT is involved in the β-oxidation of long-chain fatty acids, it also has the potential of modulating the medium-chain/long-chain acyl-CoA to acylcarnitine pool in the mitochondrial matrix.   Recent studies indicate it may be involved in providing palmitoyl residues for phospholipids of the erythrocyte membrane,[18] similar to the early studies of Bressler and Friedberg which showed carnitine enhanced the incorporation of palmitate into mitochondrial lecithin.[20]

E.   IMPORTANCE OF BUFFERING THE ACYL-COA/CoASH POOL

This area is potentially one of the most important from a metabolic point of view when one considers the carnitine deficiency states in humans, but it is also one of the least well understood and least well documented, due in part to the difficulty in obtaining direct data to prove or disprove specific hypotheses.   However, data with isolated mitochondria show that carnitine can alter specific short-chain acyl-CoA/CoASH ratios in both heart and liver mitochondria, and thereby modulate flux of acyl residues into acylcarnitines.[12,13]   It is well-established that the specific acyl-CoA to CoA ratio can be a key parameter in the short-term metabolite control of some allosteric regulated enzymes that control flux of carbon from carbohydrates and proteins into the TCA cycle.   For example, the conversion of pyruvate to acetyl-CoA by the allosteric enzyme, pyruvate dehydrogenase, and the conversion of branched-chain keto acids to branched-chain acyl-CoAs by the branched-chain keto acid dehydrogenase are affected by specific acyl-CoA to CoA ratios.   For allosteric enzymes affected by this ratio, CoASH often has an opposite overall modulating effect on the enzyme system compared to acetyl-CoA (acyl-CoA), particularly if both compounds compete for the same binding site.   Thus, the ratio of acetyl-CoA/CoASH can be more important than the absolute amount of either compound.   For example, acetyl-CoA is a (-) effector of PDH, while CoA is a (+) effector of PDH activity.   Carnitine, via the catalytic action of CAT, by altering the levels of (+) and/or (-) allosteric modifiers, has the potential of altering flux through enzymes such as pyruvate dehydrogenase and branched-chain keto acid dehydrogenase, thereby affecting catabolic metabolism.   This is more likely to occur in systems that have low amounts of carnitine or are carnitine deficient.   Little, if any, impact of high concentrations of carnitine on catabolic metabolism seems likely.[19]

F.   OTHER

The amino acid sequencce of one form of CAT,[21] one of COT,[22] and two CPTs have been determined.   Although human and rat liver CPT are very similar,[23,24] little homology exists between CAT, COT, and CPT, consistent with their different properties and functions.

III.   ACKNOWLEDGMENTS

This research has been supported by NIH grant DK 18427.   Dr. Zhi-Heng Huang kindly provided the RS-2-methylbutyryl-DL-carnitine, glutaryl-DL-carnitine, and methylglutaryl-DL-carnitine for this study, and valproyl-L-carnitine was the generous gift of Dr. Lee Carter.

IV. REFERENCES

1.   Bieber, L.L., and Farrell, S., Carnitine Acyltransferases, in <u>The Enzymes</u> XVI, 627-642, 1983.
2.   Bieber, L.L., Carnitine, <u>Ann. Rev. BCH</u> 57, 261, 1988.
3.   Fiol, C.J., and Bieber, L.L., Sigmoid Kinetics of Beef Heart CPTase: Effect of pH and malonyl CoA, <u>J. Biol. Chem.</u>, 259, 13084, 1984.
4.   Bieber, L.L., Emaus, R., Valkner, K., and Farrell, S., Possible Functions of Short-Chain and Medium-Chain Carnitine Acyltransferases, <u>Fed. Proc.</u>, 41, 2858, 1982.
5.   Bremer, J., Carnitine: Metabolism and Functions, <u>Physiol. Rev.</u>, 63, 1420, 1983.
6.   Engel, A.G., Carnitine Deficiency Syndromes and Lipid Storage Myopathies, in <u>Myology</u>, Engel, A.G. and Benker, B.Q., Eds., McGraw-Hill, 1986, chap. 57, 1663.
7.   Bieber, L.L., *Clinical Aspects of Human Carnitine Deficiency*, Borum, P.R., Ed.), Pergamon Press, New York, 1986, 102.
8.   Emaus, R.K., and Bieber, L.L, Effect of Carnitine on the Growth of the Yeast, <u>Torulopsis</u> bovina, <u>J. Biol. Chem.</u>, 258, 13160, 1983.
9.   Bieber, L.L., Sabourin, P., Fogle, P., Valkner, K., and Lutnick, R., Studies on the Formation and Utilization of Isobutyrylcarnitine, in <u>Carnitine Biosynthesis, Metabolism, and Functions</u>, Frenkel, R.A. and McGarry, J.D., Eds., Academic Press, New York, 1980, 159.
10.   Colucci, W., and Gandour, R.W., CAT: A Review of Its Biology, Enzymology and Bioorganic Chemistry, <u>Biorganic Chem.</u>, 16, 307, 1988.
11.   Farrell, S.O., Fiol, C., Reddy, J.K., and Bieber, L.L., Properties of Purified Carnitine Acyltransferases of Mouse Liver Peroxisomes, <u>J. Biol. Chem.</u>, 259, 13088-13095, 1984.
12.   Lysiak, W., Toth, P., Suelter, C., and Bieber, L.L., Quantitation of the Efflux of Acylcarnitines from Rat Heart, Brain and Liver Mitochondria, <u>J. Biol. Chem.</u>, 261, 13698, 1986.
13.   Lysiak, W., Lilly, K., DiLisa, F., Toth, P., and Bieber, L.L., Quantitation of the Effect of L-Carnitine on the Levels of Acid-Soluble Short-Chain Acyl-CoA and CoASH in Rat Heart and Liver Mitochondria, <u>J. Biol. Chem.</u>, 263, 1151, 1988.
14.   Lilly, K., and Bieber, L.L., Kinetic Characterization and DL-Aminocarnitine Inhibition of Rat Liver Carnitine Octanoyltransferase, <u>Life Sciences Intl.</u>, in press.
15.   Derrick, J.P., and Ramsay, R.R., L-Carnitine Acyltransferase in Intact Peroxisomes is Inhibited by Malonyl-CoA, <u>Biochem. J.</u>, 262, 801, 1989.

16. Lilly, K., Bugaisky, G.E., Umeda, P.K., and Bieber, L.L., The Medium-Chain Carnitine Acyltransferase Activity Associated with Rat Liver Microsomes is Malonyl-CoA Sensitive, <u>Arch. Biochem. Biophys.</u>, 280, 167, 1990.

17. Kerner, J., Froseth, J.A., Miller, E.R, and Bieber, L.L., A Study of the Acylcarnitine Content of Sows Colostrum Milk and Newborn Piglet Tissues: Demonstration of High Amounts of Isovalerylcarnitine in Colostrum and Milk, <u>J. Nutr.</u>, 14, 854, 1984.

18. Arduini, A., Mancinelli, G., and Ramsay, R.R., Palmitoylcarnitine, a Metabolic Intermediate of the Fatty Acid Incorporation Pathway in Erythrocyte Membrane Phospholipids, <u>Biochem. Biophys. Res. Commun.</u>, 173, 212, 1990.

19. Bressler, R., and Friedberg, S.J., The Effect of Carnitine on the Rate of Palmitate Incorporation into Mitochondrial Phospholipids, <u>J. Biol. Chem.</u>, 239, 1367, 1964.

20. Simi, B., Mayet, M.H., Sempore, B., and Favier, R.J., Large Variations in Skeletal Muscle Carnitine Level Fail to Modify Energy Metabolism in Excercising Rats, <u>Comp. Biochem. Physiol.</u>, 97A, 543, 1990.

21. Bloise, W., Colombo, I., Goravaglia, B., Giardini, R., Finocchiaro, G., and DiDonato, S., Purification and Properties of CAT from Human Liver, <u>Eur. J. Biochem.</u>, 189, 539, 1990.

22. Chatterjee, B., Song, C.S., Kim, J.M., and Roy, A.K., Cloning, Sequencing, and Regulation of Rat Liver COT: Transcriptional Stimulation of the Enzyme During Peroxisome Proliferation, <u>Biochem.</u>, 27, 9000, 1988.

23. Woeltje, K.F., Esser, V., Weis, B.C., Sen, A., Cox, W.F., McPhaul, M.J., Slaughter, C.A., Foster, D.W., and McGarry, S.D., Cloning, Sequence and Expression of cDNA Encoding Rat Liver Mitochondrial CPT-II, <u>J. Biol. Chem.</u> 265, 10720, 1990.

24. Finnochiaro, G., Taroni, F., Rocchi, M., Martin, A.L., Colombo, I., Tarelli, G.T., and DiDonato, S., cDNA Cloning, Sequence Analysis and Chromosomal Localization of the Gene for Human Carnitine Palmitoyltransferase, <u>Proc. Natl. Acad. Sci.</u>, 88, 661, 1991.

DISCUSSION OF THE PAPER

G. WOLDEGIORGIS *(Madison, WI)*: How do you explain that adding a malonyl CoA binding protein fraction (inactive CPT I) restores malonyl CoA sensitivity to CPT II?

BIEBER: Obviously something in the malonyl CoA binding fraction must be binding to the purified CPT-II, that results in a reduction of catalysis by CPT-II when malonyl CoA is added to the incubation media. However, I would urge caution in assuming that this is reconstitution of <u>in</u> vivo malonyl CoA sensitivity. The malonyl CoA binding fractions contain at least 6 proteins. For example, we have not even eliminated the possibility that one or more of the added proteins affects the acyl CoA binding site of the catalytic site to give an affinity for malonyl CoA, making it a competitive inhibitor. Protein/enzyme interaction could alter the catalytic behavior of CPT-II. Such possibilities must be checked before definitive conclusions can be made.

G. TREMBLAY *(Cincinnati, OH)*: Is it unrealistic to assume that the primary insult with toxicity of organic acids (eg valproate and benzoate) is the sequestration of coenzyme A resulting in the inactivation of acetyl CoA-dependent processes such as flux through pyruvate carboxylase or the ammonia-dependent stimulation of flux through the urea cycle via increased acetylglutamate, and that the action of carnitine is simply to restore normal acetyl CoA levels?

BIEBER: I think this is a realistic suggestion!

J. CARROLL *(Augusta, GA)*: Does pivaloylcarnitine lead to tissue carnitine deficiency in patients?

BIEBER: I do not know. To my knowledge this has not been determined.

E. HOPPEL *(Cleveland Hts., OH)*: Are you suggesting, with your data on unusual acylcarnitine inhibition of the exchange of radioactive carnitine with acetylcarnitine, that we should be cautious in the interpretation of acylcarnitine profiles in urines of patients analyzed by the radioexchange HPLC method?

BIEBER: Yes - We have had samples in which exchange of radioactive carnitine into acylcarnitines of urine is very slow. This could be due to the presence of inhibitory acylcarnitines. Sometimes this can be overcome by extending the incubation time to 24 hours or more. However, if the CAT preparation contains some acylcarnitine or acyl CoA hydrolase activity, the extended incubation time results in loss of acylcarnitines. Some preparations of commercial CAT contain traces of hydrolase activity!

E. BRASS *(Cleveland, OH)*: Comment: With respect to modulation of acyl-CoA ratio by carnitine, the demonstration of this effect "in vitro" require large concentrations of carnitine to be available throughout the period of observation. "In vivo" this will not be the case as the finite intracellular carnitine will rapidly reach equilibrium/steady state conditions. Thus, the acyl group flux will be limited to a short, transient period.

BIEBER: One does not need efflux of acylcarnitines to modulate the acyl CoA/CoASH ratio. The critical point is to have carnitine and carnitine acyltransferase in the compartment where the ratio is to be modified, so the enzyme system can approach equilibrium between the acyl CoA and the acylcarnitine pool.

M. NOVAK *(Miami, FL)*: The radioactive exchange method, CAT is less effective in converting medium chain acyl carnitines. Is it possible to use another carnitine acyl transferase (such as CPT) to extend this method for medium or long chain acyl carnitines?

BIEBER: Yes, we often add purified peroxisomal COT to facilitate exchange of radioactive carnitine into medium-chain acylcarnitines.

VALPROATE AND CARNITINE

Timothy P. Bohan[1], Ph.D., M.D., Peggy Rogers[2], M.D.,
and Charles R. Roe[2], M.D.

[1]Departments of Neurology, Pediatrics, and Pharmacology
University of Texas Medical School
Houston, Texas   77030

[2]Department of Pediatrics
Duke University Medical Center
Durham, NC   27710

Introduction

Valproic acid, which was discovered in the 1880's, is a branch-chained isomer of octanoic acid that is similar in structure to many of the branch-chained organic acids involved in various inborn errors of metabolism. It is an important and efficacious anticonvulsant that is useful in many different types of seizure disorders.[1] It was introduced into the clinical practice in the United States in 1978. In rare cases it can cause a severe or fatal hepatotoxicity.[2] This hepatotoxicity is characterized by microvesicular periportal steatosis and distorted mitochondria.[3]

Valproate induced hepatic changes can be reproduced in rats.[4,5] Following valproate the hepatic levels of free CoA are decreased and the hepatic levels of medium chain acyl CoA compounds are increased.[4,5] The co-administration of L-carnitine blocks the development of this valproate-induced steatosis.[6]

A valproate induced carnitine deficiency has now been reported in several studies.[7-15] The relationship of the carnitine deficiency to ammonia, the dose of valproate, and other factors has not yet been fully clarified.[6,14,16] This study was undertaken to explore the relationship between valproate therapy and the resulting carnitine deficiency.

Methods

This study was undertaken in the pediatric neurology clinic at Duke University Medical School. It was done with the approval of the committee for the protection of human subjects. It consisted of patients who were seen in the pediatric neurology clinic for seizure disorder. The patient's age, gender, seizure type, medication history, and duration of therapy were obtained from the patient chart. This study included only those patients who had received valproate alone or in combination for at least 90 days. There were a total of 38 patients between 3 and 18 years of age. Of these 24 patients were under 18 years of age and on polytherapy and 14 patients were under 18 years of age and on monotherapy. Polytherapy consisted of phenobarbital, carbamezepine, phenytoin, primidone, and/or clonazepam. Serum samples were obtained in the course of routine therapeutic monitoring. All samples were assayed by the radioenzymatic assay.

An additional patient who presented to the University of Texas Health Science Center was also included in this study. This was a 4 year old child who developed a severe hepatotoxicity after several months of valproate therapy. Laboratory analysis for plasma carnitine was also done at the Duke University Laboratory.

Results

Serum carnitine values for the patient study at Duke University are shown in Table 1. Carnitine values were markedly lower in polytherapy subjects under 10 years of age. Two between (2X2) analysis of variance revealed a significant age group relationship, $F(1,34) = 5.79$, $p = 0.022$. We also studied the relationship between carnitine levels and the valproate dose, the valproate blood level, the transaminase levels and the seizure type. No other main effects or interactions were observed.

The plasma carnitine level was also determined in a 6 and 4/12 year old Latin American female with valproate-induced hepatotoxicity. The free carnitine level was 17 and the ACYL carnitine level was 12 (see Table 2). This child, who also had Down's Syndrome, had been on numerous anticonvulsants since 2 years of age. She had been switched to valproate monotherapy 11 months prior to her presentation with fever and anorexia. Laboratory studies revealed hepatic dysfunction with a SGPT of 5,589 and a SGOT of 4,806 (normal is 7 to 40 IU/l). This child was treated with L-carnitine, 100 mg/kg/day and recovered. She was discharged to home and has had normal liver function studies for at least 3 months.

TABLE 1

The Plasma Free Carnitine Levels in Children Treated with
Valproate Monotherapy Versus Polytherapy

	Age (Years)	Free Carnitine (μMol/L)	N
Polytherapy	0-10	24.5 ± 3.7*	12
	10-18	42.9 ± 3.6	12
Monotherapy	0-10	34.5 ± 4.4	8
	10-18	38.4 ± 7.8	6
Control	All	37.0 ± 8.0	

*Different than all other values P = 0.022

TABLE 2

Plasma Carnitine Levels in Patients with Valproate-Induced
Fatal or Severe Hepatotoxicity

		Carnitine Levels (μMol/L.)				
		Patient		Control		Carnitine
Fatal Hepatic Failure	Age	Free	Acyl	Free	Acyl	Treatment
1) Bohles (1982)	3½Y	7.0	–	42.3	–	No
2) Sugimoto (1986)	13Y	19.0	6.0	44.2	8.7	No
3) Laub (1986)	3½Y	"Normal"				Yes
4) Matsuda (1986)	2Y	18.5	8.9	56.0	6.0	No*
Severe, Non-Fatal Hepatotoxicity						
5) Murphy (1985)	6M	5.3	–	50.0	–	No
6) Matsuda (1986)	7Y	39.3	31.5	56.0	6.0	Yes
7) Matsuda (1986)	2Y	16.9	7.8	56.0	6.0	No*
8) Bohan (1990)	8M	17.0	12.0	37.0	10.0	Yes

Names in the left hand column refer to the appropriate
reference except for Bohan which is a personal communication.

*These two patients are twins

Discussion

These data indicate that young age and polytherapy are the primary risk factors for valproate-induced carnitine deficiency. Unlike the original report of valproate-induced carnitine deficiency which suggested a dose response relationship,[7] our data did not support such a conclusion nor was it observed by Laub.[14] A recent review (Table 3) indicated that most studies did not compare the carnitine values in monotherapy versus polytherapy patients. In the two studies that did such a comparison, polytherapy patients had lower levels. This is in agreement with our observation that polytherapy produces a greater carnitine deficiency than

TABLE 3

The Free and Acyl Carnitine Levels (μMol/L) in Patients Treated with Valproate and/or Other Antiepileptic Drugs
A Summary of Eight Studies

| Study | N | Rx | Valproate | | Other AED | | Control | |
			Free	Acyl	Free	Acyl	Free	Acyl
Ohtani (1982)	14	M+P	28.6	4.0	43.0	5.6	44.2	5.8
Murphy (1985)	13	M+P	39.5		50			
Matsuda (1986)*	5	P	19.4	12.7			56.0	6.0
Morita (1986)	12	M+P	21.5	12.0	31.5	12.0	51.7	9.7
Melegh (1987)	11	M	16.8	7.5			26.5	8.4
Melegh (1989)	10	M+P	26.1	6.5			42.7	6.1
Rodriguez (1989)	34	M+P	26.4	10.5	41.2	6.9		
Laub (1986)	14	M	34.7	11.7	40	7	41.5	8
	7	P	28.9	7.5				
Komatsu (1987)	8	M	55.7		48.5		57.3	
		P	42.5		52.8			
Bohan (1990)	16	M	36.2				37	10
	27	P	33.7					

Rx = Type of therapy   M = Monotherapy   P = Polytherapy   M+P = A mixture of monotherapy and polytherapy

*Three of these five patients had a Reye's-like syndrome and two died.

monotherapy. Hence, young age and polytherapy appear to be the primary risk factors for both valproate-induced carnitine deficiency and valproate-induced fatal hepatic failure.[2]

Although there are numerous case reports and case series in the literature documenting valproate-induced hepatic failure, there are very few reports of the carnitine levels in patients with severe or fatal valproate-induced hepatoxicity. The studies that are available (eight patients) are summarized in Table 2. Four patients had fatal hepatic failure and 4 had severe but nonfatal hepatotoxicity. Of these 8 patients only 1 had a normal level of carnitine. This patient also had minimal evidence of liver dysfunction. Hence, what data is available indicates that patients with fatal or severe valproate hepatotoxicity had an associated carnitine deficiency. The role of carnitine supplementation once severe toxicity has developed is not yet clear.

Future studies in this field should focus on determining the carnitine levels in patients with severe or fatal valproate-induced hepatotoxicity. Such a study could be facilitated by an international registry. These patients should have not only carnitine determinations, but also, a full metabolic workup since adverse reactions to valproate may be related to occult inborn errors of metabolism. Due to the low incidence of valproate-induced hepatoxicity, it would not be practical to prospectively determine the effect of carnitine supplementation on valproate-induced hepatotoxicity.

Bibliography

1. Bourgeois BFD. Valproate: Clinical use. Antiepileptic Drugs. 3rd ed. Levy R, Mattson R, Meldrum B, Penry JK, Dreifuss FE, eds. New York: Raven Press, 1989; 633-642.

2. Dreifuss FE, Santilli N, Langer DH, et al. Valproic acid hepatic fatalities: A retrospective review. Neurology 1987; 37:379-385.

3. Zimmerman H, Ishak K. Valproate-induced hepatic injury; Analyses of 23 fatal cases. Hepatology 1982; 2:591-597.

4. Lewis JH, Zimmerman HY, Garrett CT, Rosenberg E. Valproate-induced hepatic steatogenesis in rats. Hepatology 1982; 2(6):870-873.

5. Olson MJ, Handler JA, Thurman RG. Mechanism of zone-specific hepatic steatosis caused by valproate: Inhibition of ketogenesis in periportal regions of the liver lobule. Molecular Pharmacology 1986; 30:520-525.

6. Sugimoto T. Reye-like syndrome associated with valproic acid. Brain and Development 1983; 5:334-337.

7. Ohtani Y, Matsuda I. Valproate treatment and carnitine deficiency. Neurology 1984; 34:1128-1129.

8. Murphy JV, Marquardt KM, Shug A. Valproic acid-associated abnormalities of carnitine metabolism. Lancet 1985; 1:820-821. Letter.

9. Matsuda I, Ohtani Y, Ninomiya N. Renal handling of carnitine in children with carnitine deficiency and hyperammonemia associated with valproate therapy. J Pediatr 1986; 109:131-134.

10. Morita J, Yuge K, Yoshino M. Hypocarnitinemia in the handicapped individual who receives a polypharmacy of antiepileptic drugs. Neuropediatr 1986; 17:203-205.

11. Melegh B, Kerner J, Kispal G, Acsadi G, Dani M. Effect of chronic valproic acid treatment on plasma and urine carnitine levels in children: Decreased urinary excretion. Acta Paediatrica Hungarica 1987; 28(2):175-178.

12. Melegh B, Kerner J, Acsadi G, Lakatos J, Sandor A. L-carnitine replacement therapy in chronic valproate treatment. Neuropediatrics 1990; 21:40-43.

13. Rodriguez-Segade S, Alonso de la Pena C, Tutor J, et al. Carnitine deficiency associated with anticonvulsant therapy. J Clin Chemica Acta 1989; 181:175-182.

14. Laub MC, Paetzke-Brunner I, Jaeger G. Serum canitine during valproic acid therapy. Epilepsia 1986; 27:559-562.

15. Komatsu M, Kodama S, Yokoyama S, et al. Valproate-associated hyperammonemia and DL-Carnitine supplement. Kobe J Med Sci 1987; 33:81-87.

16. Coulter D. Carnitine, valproate, and toxicity. J. Child Neurology 1991; 6:7-14.

DISCUSSION OF THE PAPER

M. NOVAK *(Miami, FL)*: Is there any direct evidence of decreased beta oxidations in patients on valproic acid?

BOHAN: In a recent study of three patients with valproate induced hepatotoxicity, there was evidence of impaired valproic acid beta oxidation[1]. Those authors reviewed other cases for which biochemical data are available and concluded that in all published cases of valproate induced hepatotoxicity there was impaired beta oxidation. The same group of authors published data on the metabolism of valproic acid during long-term therapy[2]. There was a tendency for the proportion of the drug cleared by beta oxidation to decrease as the drug dose increased, while the proportion cleared by direct glucuronide conjugation tended to increase. Hence, there is data with valproyl hepatotoxicity of decreased beta oxidation and a suggestion that patients on high dose chronic therapy may have a mild decrease in beta oxidation.

G. TREMBLAY *(Kingston, RI)*: Do you suppose some of the variation that you see in the adverse reaction to valproate may be a function of variation in the capacity of patients to conjugate valproate with glucuronic acid.

BOHAN: Improved rate of conjugation would have the same protective action as carnitine. Several studies have reported on valproate metabolism during severe hepatotoxicity. Only one study reported glucuronide metabolites[3]. In two of their three patients, there was increased valproic acid-glucuronide formation and excretion. In their third patient who had renal insufficiency, there may have been diminished valproic acid-glucuronide excretion. Hence to date, there is no evidence that the adverse reaction of valproate may be a function of variation in the capacity of patients to conjugate valproate with glucuronic acid. Since two of the three patients to conjugate valproate with glucuronic acid. Since two of the three patients had improved rates of conjugation and yet severe hepatotoxicity, it would appear that increased conjugation did not have protective action.

I. FRITZ *(Canada M5G 126)*: How does valporic acid work in the brain? If valproic acid administration has such potentially dire hepatotoxic effects when given to young patients, why is it not preferable to treat with other anticonvulsants?

BOHAN: Valproic acid is becoming one of the most widely used anticonvulsants in children with seizures because it is efficacious in so many patients with so many different kinds of seizures. Furthermore, it has the least effects on behavior and cognition. Hence, the combination of increased efficacy and reduced cognitive side effects has made valproate the mainstay of pediatric epileptologists. When appropriate, less toxic anticonvulsants are used. Although the hepatotoxic effects can be dire, they are relatively rare and if caught early enough often, but certainly not always, are reversible.

I. FRITZ *(Canada M5G 126)*: If the major interest is to determine the mechanism by which carnitine alleviates or prevents some of the symptoms of valproate induced hepatotoxicity, than surely experimental animal model would be more appropriate to investigate then patients having side effects on valproate therapy! What is the best animal model thus far available?
The pathological effects observed more often in young children than in adults given valproic acid suggests that adequate experimental animal models should include the use of neonatal or immature mice or rats.

BOHAN: There have been numerous studies of the adverse effects of valproic acid in animals. Most have focused on rats or mice. It is unclear whether one animal model is superior to others. Certainly in these models the coadministration of carnitine alleviates the pathological changes which are induced by very high toxic doses of valproate. What is important now is that this animal data be carried to the bedside to see if it is relevant in children as well as in animals.

1.   (Eadie et al, 1990) Drug Quarterly Journal of Medicine. 77: 1229-40, 1990.
2.   (Dickinson et al, 1989) Therapeutic Drug Monitoring. 11:127-133, 1989.
3.   (Eadie et al, 1990) Drug Quarterly Journal of Medicine. 77:1229-40, 1990.
4.   Eadie, M.J., McKinnon, G.E., Dunstan, P.R., Maclaughlin, D., and Dickinson, R.G. Valproate Metabolism During Hepatotoxicity Associate with the Drug.

BIOCHEMICAL PROPERTIES OF PROPIONYL- AND ISOVALERYL-CARNITINE

Di Lisa F., Menabò R., Barbato R., Miotto G., Venerando R. and Siliprandi N.
Dept. of Biological Chemistry, University of Padua and Centro per lo Studio della Fisiologia dei Mitocondri, Padova, Italy

I. INTRODUCTION

In normal human subjects propionyl- and isovaleryl carnitine are present in biological fluids in very minute amounts. In propionic and isovaleric aciduria these esters become a preponderant part of the short chain acyl carnitine fractions in both plasma and urine and their concentrations sharply increase upon carnitine administration.[1,2] Short-chain acyl carnitines are easily transported across mitochondrial membranes by a translocase system, seemingly responsible for the exchange transport of both long- and short-chain acyl carnitine.[3] The successive diffusion of short-chain acyl carnitine across cell membranes and the renal barrier accounts for the presence of these esters in blood and urine. The short-chain acyl carnitine transport is reversible both across mitochondria[3] and plasma membranes.[4] Therefore it appeared a priory reasonable to assume that administered short-chain acyl carnitine could easily reach the mitochondrial matrix space. Here their transformation into the corresponding acyl-CoA preludes the integration of the acyl moieties in mitochondrial metabolic pathways.

In the present paper the effects of propionyl carnitine on contractility and energy-linked processes of perfused rat heart will be reported, as well as those of isovaleryl carnitine on liver proteolysis and calpains activity. The meaning of the action of these short-chain carnitines and the relative mechanisms will also be discussed.

II. PROPIONYL CARNITINE

A. General

In animals, the propionyl group of propionyl-CoA derives from three sources: 1) activation of propionate, present in high concentrations in ruminants, but also in detectable amounts in human tissues; 2) β-oxidation of odd fatty acids and the side chain of cholesterol; 3) catabolism of isoleucine, valine, threonine and methionine, the latter both via α-ketobutyrate and the recently discovered 3-methylthiopropionyl- CoA.[5]

The major metabolic pathway from propionyl-CoA through succinate leads into the tricarboxylic acid cycle in an "anaplerotic" (filling up) process.[6] This pathway is interrupted in two inborn diseases: <u>propionic and methylmalonic aciduria</u> [1,2] caused by an abnormally low, or missing, activity of propionyl-CoA carboxylase (dependent on biotine) and methylmalonyl-CoA mutase (dependent on 5'-deoxyadenosylcobalamin) respectively.

In a second pathway, detected in individuals with propionic aciduria,[7] propionyl-CoA leads to the formation of acetyl-CoA, via malonic semialdehyde.

Furthermore both in propionic and methylmalonic aciduria 2-methylcitrate is formed by the condensation of propionyl CoA and oxaloacetate.[7]

B. Transport and metabolic action.

The above mentioned processes take place within the mitochondrial matrix space and the only possibility for the acyl moieties to be released is either hydrolysis of the acyl CoAs, mediated by the still poorly studied deacylases,[8] or conversion of acyl CoAs into the corresponding acyl carnitine. The latter process is catalyzed by acetyl carnitine transferase , which, despite its internationally accepted name (CAT,EC 2.3.1.7), is most active towards propionyl carnitine.[3] This transferase is present in liver peroxisomes and in mitochondria, especially those from heart and kidney.[9] In tandem with carnitine translocase, CAT allows a two-way movement of acyls across mitochondrial membranes.[3] An important consequence of this transport is the homeostasis of the propionyl CoA/CoA ratio. When the outward movement is retarded, generally owing to the scarce availability of free carnitine (carnitine insufficiency,[10]) propionyl-CoA accumulates in mitochondria. The consequent metabolic derangements are due both to the concurrent shortage of free CoA and the inhibitory action of accumulated propionyl-CoA on some enzymatic activity. Namely propionyl-CoA inhibits the formation of succinate from succinyl-CoA, via allosteric inhibition of succinate:CoA ligase[11] and ureagenesis via the inhibition of carbamoylphosphate synthetase.[12]

More precisely these metabolic alterations are the result of an unbalance between the rate of formation of propionyl-CoA (from propionate either formed endogenously or transported from outside) and that of its elimination (either via metabolism or propionyl carnitine formation). This is illustrated by the results obtained both on isolated mitochondria and perfused rat heart.

TABLE I
ACTION OF PROPIONATE AND PROPIONYL CARNITINE ON OLEATE OXIDATION IN RAT LIVER MITOCHONDRIA

Additions	CO_2	ATP	ADP	AMP
	(CPM x mg^{-1} x 5 min^{-1})	(nmoles x mg. prot.$^{-1}$)		
^{14}C-Oleate	4600	4.3	5.5	1.1
" + Propionate	990	1.4	5.1	4.1
" + Propionylcarnitine	8900	5.3	6.2	1.5

The incubation mixture contained in a final volume of 1 ml 15 mM phosphate buffer pH 7.4, 20 mM NaCl, 50 mM KCl, 5 mM MgCl$_2$, 25 mM sucrose, 0.1 mM ^{14}C-oleate (U) (2.1x10^6 cpm/µmole), 8 mg mitochondrial protein. Temperature 25°C; time 5 minutes. ^{14}CO$_2$ was trapped by 0.2 ml of 1 M hyamine hydroxide. Both hyamine^{14}CO$_2$ and the HClO$_4$ soluble activity were determined by liquid scintillation counting. The acid soluble activity represents the metabolites derived from the acid-insoluble oleate. The concentration of ATP, ADP and AMP in mitochondria were measured enzymatically. 28

When added to rat liver mitochondria, propionate decreases the concentration of endogenous ATP and concomitantly increases that of AMP,[13] as is expected from its activation into propionyl-CoA.[14] Conversely propionyl carnitine, being transformed into propionyl-CoA without any energy expense, does not significantly modify the adenine nucleotide pool.

The inhibition of oleate oxidation by propionate and its stimulation by propionyl carnitine are very likely the consequences of the opposite effects of the two compounds on mitochodrial ATP. The shortage of ATP caused by propionate, together with the inhibition of the ATP dependent thiokinase (acid:CoA ligase ATP) by the concomitantly accumulated AMP[15] may account for the inhibition of oleate oxidation. Conversely the stimulation of this process by propionyl carnitine might depend both on the accelerated flux in the Krebs cycle, promoted by the anaplerotic effect of the newly formed propionyl-CoA and on the larger availability of intramitochondrial carnitine derived from propionyl carnitine in the CAT-catalyzed reaction.

In isolated perfused rat hearts addition of propionate to the perfusate induced the disappearance of free CoA and the concomitant formation of both propionyl- and methylmalonyl-CoA (Fig.1A) and that of a substantial amount of propionyl carnitine at expenses of free carnitine

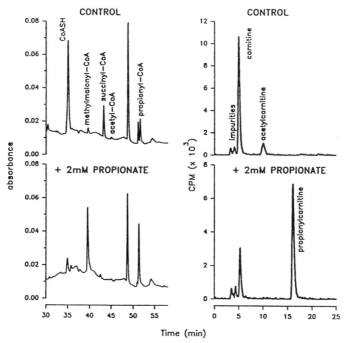

Figure 1 - CoA and carnitine fractions of isolated rat hearts perfused under aerobic conditions for 60 min. in the presence of 11 mM glucose as a substrate. At the end of perfusion, the hearts were freeze clamped and the frozen tissues were deproteinized with 0.85 N PCA (1:5 v/w) + 1 mM DTT. 0.1 ml of neutralized PCA extract was used for CoA analysis as described. 16 For carnitine measurement, the neutralized PCA extract was passed through an anion exchange column (Dowex 1) and the effluent was utilized to exchange [³H]carnitine into the acylcarnitine of the sample as described.29 [³H]acylcarnitine were separated by a reverse-phase column and the radioactivity in the effluent was monitored by an on-line flow β–detector. Each peak was quantitated on the basis of the total carnitine of the sample assayed separately with a radioenzymatic procedure.30

Remarkably, in spite of these rather drastic changes of mitochondrial metabolism, under aerobic conditions propionate did not induced appreciable modifications in contractility efficiency (results not reported).

Conversely by replacing propionyl carnitine for propionate in the perfusate, propionyl-CoA, otherwise undetectable, was formed, but not methylmalonyl-CoA (results not reported). This difference and the preservation of free CoA suggests a slower and more controlled access rate of propionyl carnitine into mitochondria. It is also conceivable that in perfused heart, as observed in isolated rat liver mitochondria,[13] endogenous ATP is differently affected by propionate and propionyl carnitine.

Conceivably the different metabolic effects of propionate and propionyl carnitine are the result of both a different rate of transport of the two species into the inner mitochondrial space and the distinct pathway for their transformation into propionyl-CoA.

C. Cardiac contractility and mitochondrial energy linked processes.

The contractility of isolated perfused rat heart submitted to a transient ischemia, followed by reperfusion was significantly impaired.

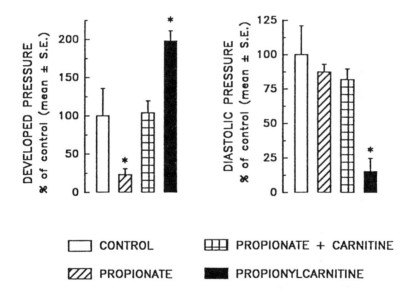

Figure 2 - Contractile activity of isolated hearts after 90 min of ischemia (0.2 ml/min) and 20 min of reperfusion expressed as a percentage of the control group (perfused only with glucose). Developed and diastolic pressures were monitored by means of a fluid filled latex balloon inserted into the left ventricle and connected to a pressure transducer. Under aerobic conditions, irrespective of the composition of the perfusion medium, the developed pressure was 75 ± 5 mmHg and the diastolic pressure was set to 10 ± 2 mmHg by a gradual filling of the latex balloon. Values are the means $\pm$ S.E. of at least 6 experiments. * = $p < 0.05$.

This damage was further aggravated by addition of propionate to the perfusate. Indeed, as shown in Fig 2, addition of propionate to the perfusate significantly reduced the systolic pressure, which was, however preserved if propionate was added together with carnitine, indicating that the severe reduction of free carnitine (see Fig.1) is relevant for myocardial contractility. Conversely addition of equimolar amounts of propionyl carnitine resulted in a clear cut improvement of contractility, as reflected by the significant increase of the systolic pressure.

These dynamic changes are reflected by the energy linked processes (transmembrane potential and oxidative phosphorylation efficiency) observed in mitochondria isolated either from aerobically perfused hearts (normoxic) or from "ischemic" hearts. (Table II) In the latter mitochondria transmembrane potential was significantly decreased and upon addition of ADP its recovery was much slower than in "normoxic" mitochondria, indicating a lower efficiency of oxidative phosphorylation. In "ischemic" mitochondria transmembrane potential was significantly impaired by propionyl carnitine. (Table II)

TABLE II
EFFECT OF PROPIONYL CARNITINE ON OXIDATIVE PHOSPHORYLATION EFFICIENCY OF MITOCHONDRIA FROM ISCHEMIC RAT HEARTS

| | Transmembrane Potential ($\Delta\Psi$) mV | | |
	Steady state	after ADP	Recovery
Normoxia	163(4)	129(3)	163(3)
Ischemia	143(5) *	121(4)	132(4) *
Ischemia + Propionylcarnitine	154(4)	123(4)	154(4)

After completion of either a normal flow (12ml/min) "normoxia" or a low flow (1ml/min) "ischemia" perfusion (60 min), rat hearts were homogenized in a medium containing 0.18 M KCl, 0.01 M EDTA, 0.5% albumin,10 mM Tris, pH 7.4 and mitochondria were isolated by differential centrifugation.[31] Membrane potential ($\Delta\Psi$) was measured with tetraphenylphosphonium (TPP+) selective electrode according to Kamo et al [32]. Mitochondria incubation was carried out at 20°C with 0.5 mg mitochondrial protein/ml in the following standard medium: 0.2 sucrose, 100 mMHeps, pH 7.4, 5 mM succinate, 1.25 µM rotenone. Incubation volume 3 ml. The concentration of propionyl carnitine added to the perfusate was 1.5 mM; that of ADP added to isolated mitochondria 0.25 mM . Addition of ADP ("after ADP") involved a transient utilization of part of the transmembrane potential for ADP phosphorylation, accounting for the decrease of $\Delta\Psi$ in "steady state" conditions. At the completion of ADP phosphorylation $\Delta\Psi$ reached again the original steady state value ("recovery").

These results indicate that: 1) the energy produced by mitochondria is relevant, if not critical, for heart contractility and 2) the different effects of propionate and propionyl carnitine on cardiac contractility appear to be direct consequences of their action at the mitochondrial level.[16]

In this regard it is worthy to outline again that, unlike propionyl carnitine, propionate causes a drainage of ATP and a depletion of free CoA. The latter may be critical for the utilization of both propionate and α-ketoglutarate, hence for the flux in the Krebs cycle. Under hypoxic conditions these changes are relevant, if not entirely responsible, for the decreased efficiency of cardiac contractility, as well as for the impairment of mitochondrial trasmembrane potential.

III. ISOVALERYL CARNITINE

A. Inhibition of lysosomal proteolysis.

Isovaleryl carnitine, a leucine catabolite, is produced in substantial amounts by rat liver perfused with carnitine plus α-ketoisocaproate.[17] Both α-ketoisocasproate and its parent amino acid, l-leucine, are known to inhibit lysosomal proteolysis in rat liver perfused in the absence of amino acids.[18] This raises the question as to whether leucine is responsible for the inhibition per se, or via some of its catabolites. Among these isovaleryl carnitine, owing to its capability to be transported across cellular and intracellular membranes seemed to us a suitable catabolic intermediate for this study.

Surprisingly, the addition of isovaleryl carnitine to the perfusate in concentrations from 0.1 to 1 mM inhibited liver proteolysis induced by amino acid deprivation as efficiently as equimolar amounts of leucine.[19] Also the typical concentration-response curve previously found for leucine[20] was mimicked by this ester. The inhibition of proteolysis by isovaleryl carnitine is a process that certainly does not involve leucine, since the oxidative decarboxylation of α-ketoisocaproate into isovaleryl-CoA, the immediate precursor of isovaleryl carnitine, is an irreversible reaction. One possibility is that the isovaleryl group, common to both leucine and isovaleryl carnitine might be the signal recognizable by a receptor regulating lysosomal proteolysis.

The inhibitory action of isovaleryl carnitine is remarkably specific, since the isovaleryl carnitine related compounds so far tested, exhibited a much lower effect, or none at all.

The site of inhibition of isovaleryl carnitine seems to be at the level of the vacuolar-lysosomal system (unpublished data).

Since only negligible amount of added isovaleryl carnitine is detectable within liver cells, either in perfused liver or in isolated hepatocytes, it is conceivable that the relative receptor is located on the plasma membrane. It would be otherwise difficult to reconcile the antiproteolytic effect of isovaleryl carnitine with its absence in the cells.

B. Activation of calpains.

Successively we address the problem as to whether isovaleryl-l-carnitine might also modify some of the cytosolic proteolytic systems and particularly that of calpains, ubiquitous Ca^{2+} dependent proteases present in all mammalian cells and subject to regulation by a variety of mechanisms.[21] Calpains require relatively high, non physiological concentrations of Ca^{2+} for activity. This, together with the presence of an endogenous inhibitor,[21] insures that the proteolytic activity is not expressed in an uncontrolled manner.

To our surprise isovaleryl-l-carnitine, rather than an inhibitor resulted a potent activator of calpain of human neutrophils.[22] The D isomer of isovaleryl carnitine and both L-isobutyryl- and L-methylbutyril carnitine, derivatives of valine and isoleucine respectively, were almost ineffective.[22] The activating action of isovaleryl carnitine on calpain has been envisaged in a remarkable increase of affinity of calpain for Ca^{2+}. Furthermore it was shown that isovaleryl carnitine acts synergistically with the cytoskeleton activator and that the two activators bind to different sites of the calpain molecule.[22]

Calpain from human neutrophils[23] as well as that from human erythrocytes[24] contain only the calpain I isoform whereas most mammalian tissues contain two distinct isoenzymes: calpain I, requiring 5-50 µM Ca^{2+} and calpain II 0.2-0.5 mM Ca^{2+} for half maximal activity.[25] It was found that isovaleryl carnitine exerts a potent activating effect specifically directed on calpain II. In the presence of 1 mM isovaleryl carnitine the activity of this class of calpains is expressed at Ca^{2+} concentrations one order of magnitude lower.[26] Furthermore isovaleryl carnitine and phospholipid vesicles, the latter capable of activating per se calpain activity, exhibit an additive effect at µM Ca^{2+} concentrations, at which this calpain is otherwise inactive.

The meaning of this unsuspected action of isovaleryl carnitine on calpain remains obscure. For the moment being it is tempting to speculate that the activation of calpains by isovaleryl carnitine might provide a mechanism for the initiation of controlled proteolysis involving leucine catabolites, when they reach a critical concentration in the cell. In principle this may happen either in prolonged fasting and exertion, either conditions leading to an abnormal leucine utilization.

The circumstance that isovaleryl carnitine inhibits lysosomal proteolysis and stimulates the limited Ca^{2+}-dependent limited proteolysis, mediated by the calpains, appears as a paradox. Hypothetically isovaleryl carnitine might initiate a process of limited proteolysis in order to promote a remodelling rather than a degradation of cytosolic proteins. For such a purpose the activation of calpains and the inhibition of lysosomal proteases are equally well suited. Finally, since the proteins most susceptible to the calpain attack are those of the cytoskeleton fibers, which make extensive connections to mitochondria and other organelles[27] it is also conceivable that cytoskeleton disconnection promoted by calpain action could render lysosomal proteases less accessible to the target proteins. Through the mechanisms herein discussed isovaleryl carnitine might represent a proper signal for initiating these structural and functional modifications.

REFERENCES

1. Roe, C. R., Millington, D. S., Maltby, D. A., Kahker, S. G. and Bohan, T. P., L-carnitine therapy in isovaleric acidemia, J.Clin.Invest., 74, 2290, 1984.
2. Duran, M., Keting, D., Beckeringh, T. E., Leupold, D. and Wadman, S. K., Direct identification of propionyl carnitine in propionic acidemia: biochemical and clinical results of oral carnitine supplementation, J.Inher.Metab.Dis., 9, 202, 1986.
3. Bieber, L. L., Carnitine, Ann.Rev.Biochem., 57, 261,1988.
4. Siliprandi, N., Sartorelli, L., Ciman, M. and Di Lisa, F., Carnitine: metabolism and clinical chemistry, Clin.Chim.Acta., 183, 13, 1989.
5. Scislowki, W. D., Hokland, B .M. and Davies van Thienen, W. I. A., Methionine metabolism by rat muscle and other tissues, Biochem.J., 247, 35, 1987.
6. Lehninger, A. L., Principles of Biochemistry, Worth Publ., New York.,1982, 969.
7. Przyrembel, H., Bremer, H. J. and Duran, M., Propionyl-CoA carboxylase deficiency with overflow of metabolites of isoleucine at all levels, Eur.J.Pediatr., 130, 1, 1979
8. Lee, S. H. C. and Davies, E. J., Amino acid catabolism by perfused rat hindquarter. The metabolic fates of valine, Biochem.J., 233, 621, 1986.

9.   Ferri, L., Valente, M., Ursini F., Gregolin, C. and Siliprandi N., Acetyl-carnitine formation and pyruvate oxidation in mitochondria from different rat tissues, Bull.Mol.Biol.Med., 6, 16, 1981.

10.   Chalmers, R. A., Roe, C. R., Stacey, T. E. and Hoppel, C. L., Urinary excretion of L-carnitine and acylcarnitines by patients with disorders of organic metabolism: evidence for secondary insufficiency of L-carnitine, Pediatric Res., 18, 1325, 1984.

11.   Stumpf, D. A., McFee, J., Parks, J.K. and Eguren,L., Propionate inhibition of succinate:CoA ligase (GDP) and the citric acid cycle in mitochondria, Pediatric Res., 14, 1127, 1980.

12.   Gruskay, J. A. and Rosemberg, L. E., Inhibition of hepatic mitochondrial carbamoyl phosphate synthetase (CPSI) by acyl CoA esters: possible mechanisms of hyperammonemia in the organic acidemias, Pediatric Res., 13, 475, 1979.

13.   Ciman, M., Rossi, C. R. and Siliprandi, N., On the mechanism of the antiketogenic action of propionate and succinate in isolated rat liver mitochondria, FEBS Letters, 22, 8, 1972

14.   Scholte, H. R., Wit-Peeters, E. M. and Bakkers, J. C., The intracellular and intramitochondrial distribution of short-chain acyl CoA synthetase in guinea pig heart, Bioch.Biophys.Acta, 231, 479, 1971.

15.   Rossi, C. R., Alexandre, A., Carignani, G. and Siliprandi,N., The role of mitochondria adenine nucleotide pool on the regulation of fatty acid and α-ketoglutarate oxidation, Advances in Enzyme Regul., 10, 171, 1972.

16.   Di Lisa, F., Menabò, R. and Siliprandi, N., L-Propionyl-carnitine protection of mitochondria in ischemic rat hearts, Mol.Cell.Biochem., 88, 169, 1989.

17   Hokland, B. N., Uptake, metabolism and release of carnitine and acyl carnitine in the perfused rat liver, Bioch.Biophys.Acta, 961, 234, 1988.

18.   Mortimore, G. E. and Poso, A. R., Intracellular protein catabolism and its control during nutrient deprivation and supply, Ann.Rev.Nutr., 7, 39, 1987.

19.   Miotto, G., Venerando,R. and Siliprandi, N., Inhibitory action of isovaleryl-L-carnitine on proteolysis in perfused rat liver, Bioch.Biphys.Res.Comm., 158, 797, 1989.

20.   Poso, A. R., Wert, J. J. and Mortimore, G. E., Multifunctional control by amino acids on deprivation induced proteolysis in liver:role of leucine, J.Biol.Chem., 257, 12114, 1982.

21.   Pontremoli, S. and Melloni, E., Extralysosomal protein degradation, Ann.Rev.Biochem., 55, 455, 1986.

22.   Pontremoli, S., Melloni, E., Michetti, M., Sparatore, B., Salamini, F., Siliprandi,N. and Horecker, B. L., Isovalerylcarnitine is a specific activator of calpain of human neytrophils, Bioch.Biophys.Res.Comm., 148, 1189, 1987.

23.   Pontremoli, S., Sparatore, B., Salamino, F., Michetti, M., Sacco, O. and Melloni, E., Reversible activation of human neutrophils calpain promoted by interaction with plasma membrane, Biochem.Inter., 11, 35, 1986.

24.   Pontremoli, S., Melloni, E., Salamino, F.,Sparatore, B., Michetti, M., Sacco, O. and Bianchi, G., Characterization of the defective calpain-endogenous calpain inhibitor system in erythrocytes from Milan hypertensive rats, Bioch.Biophys.Res Comm., 139, 341, 1986.

25. Mellgren, R. L., Calcium dependent proteases: an enzyme system active at cellular membranes?, FASEB J., 1, 110, 1987.

26. Pontremoli, S., Melloni, E., Viotti, P. L., Michetti, M., Di Lisa,F. and Siliprandi,N., Isovalerylcarnitine is a specific activator of the high calcium requiring calpain forms, Bioch.Biophys.Res.Comm., 167, 373, 1990.

27. Lin, A., Krockmalnic, G. and Shelton, P., Imaging cytoskeleton-mitochondrial membrane attachments by embedment-free electron microscopy of saponin-extracted cells, Proc.Natl.Acad.Sci., 67, 8565, 1990.

28. Drahota, Z., Alexandre, A., Rossi, C. R. and Siliprandi, N., Organization and regulation of fatty acid oxidation in mitochondria of brown adipose tissue, Bioch.Biophys.Acta, 205, 491, 1970

29. Bieber, L. L. and Kerner, J. K., Short chain acyl carnitines: identification and quantitation, Meth. Enzymol., 123, 264, 1986.

30. Bieber, L. L. and Lewin, L. M., Meeasurement of carnitine and acylcarnitines, Meth. Enzymol., 72, 276, 1981.

31. Lindenmayer, G. E., Sordahl, L. A. and Schwartz, A., Reevaluation of oxidative phosphorilation in cardiac mitochondria from normal animals and animals in heart failure, Circ. Res., 23, 439, 1968.

32. Kamo, N., Muratsugu, M., Hongoh, R. and Kobatake, Y., Menbrane potential of mitochondria measured with an electrode sensitive to tetraphenyl phosphonium and relationship between proton electrochemical potential and phosphorylation potential in steady state, J.Membr.Biol., 49, 105, 1979

DISCUSSION OF THE PAPER

G. LOPASCHUK *(Canada)*: Did you preload your isolated perfused hearts with carnitine? Could the difference between carnitine and proprionylcarnitine in protection of the ischemic heart be due to the amount of carnitine that actually got into the heart; was proproinylcarnitine simple facilitating uptake of carnitine by the heart?

F. DI LISA: All the substances, i.e., either propionate, carnitine or propionylcarnitine were added to the media during the entire perfusion periods. The fact that carnitine was able to prevent the damage induced by propionate suggests that even after a short time exposure a significant import of carnitine can occur into myocardial cells. However, it is very difficult to obtain a reliable measure of intracellular carnitine upon carnitine perfusion. In fact little or no data are available concerning the distribution of carnitine between vascular, interstitial and intracellular compartments. On the other hand, a washout procedure could result in the release of the accumulated carnitine. These problems obviously do not affect the tissue carnitine assay upon propionylcarnitine perfusion. The ability of propionylcarnitine to increase the free carnitine content and its relationship with the degree of myocardial protection will be the subject of further investigation in our laboratory.

F. HOMMES: Carnitine is used in the treatment of isovaleric acidemia especially in the acute phase. This will undoubtedly lead to an increase in intracellular isovaleryl carnitine in this condition.   Although the effects of isovaleryl carnitine on calpain activation are not yet completely clear, it is fair to say that the presence of high levels of isovaleryl carnitine will upset the normal physiological control mechanisms.   What effect will your findings have on the treatment of isovaleryl acidemia patients with carnitine?

F. Di LISA: We are studying the changes in calpain activity in the erythrocytes and neutrophils in a case of isovaleric acidemia during treatment either with carnitine or glycine.   Upon carnitine administration (200 mg/kg/day), the isovalerylcarnitine plasma concentrations of this patient (20 nmoles/ml, unpublished results) are much lower those required for calpain activiation in our experimental conditions.   However an accumulation of IVC by erythrocytes or neutrophils cannot be excluded since the esterified/free carnitine ratio in these cells has been reported to be higher than in plasma[1].

J. CHRISTODOULOU (Canada): Is valproate-induced pancreatitis related to alterations in the acyl CoA: CoASH ratio or to some other factor?

F. Di LISA: I'm not familiar with pancreas pathology, but as far as valproate toxicity is concerned, I don't think it can be entirely explained by a reduction of CoASH or valproyl-CoA and valproylcarnitine formation.   Preliminary data obtained in our laboratory showed that the addition of valproate to liver, heart or kidney mitochondria induced the formation of appreciable amounts of valproyl-CoA only in liver.   We never detected the appearance of valproylcarnitine and even in liver only a minor decrease of CoASH occurred.   Other (additional) factors have to be taken into account to fully understand valproate effects.   Several reports indicate that plasma acctylcarnitine is acultely increased after valproate administration[2].   Thus if CoASH is reduced, this could be due to an increased formation of acetyl-CoA which in turn would lead to an increased release of acetylcarnitine.   It's temping to speculate that valproate inhibits acetyl-CoA utilization by impairing either the Krebs cycle or the respiratory chain.   The recent report of a direct inhibition of valproate on the α-oxoglutarate dehydrogenase of brain mitochondria[3] further supports this mechanism of action.

1.   Maccari, F., Hülsmann, W.C., (Acyl) carnitine distribution between plasma, erythrocytes and leukocytes in human blood, Clin. Chem., 35, 711, 1989.

2.   Rozas, I., Camina, M.F., Paz, J.M., Alonso, C., Castro-Gago, M., Rodriquez-Segade, S.   Effects of acute valproate administration on carnitine metabolism in mouse serum and tissues, Biochem. Pharmacol., 39, 181-185, 1990.

3.   Ludes, A.S., Parks, J.K., Freeman, F., Parker, W.D. (1990) Inactivation of beef α-keto dohydrogenase complex by valproic acid metabolites. Possible mechanism of anticonvulsivant and toxic actions J. Clin. Invest. 86, 1574-1581, 1990.

METABOLIC FATE OF DIETARY CARNITINE IN HUMANS[*]

Charles J. Rebouche
Department of Pediatrics
The University of Iowa College of Medicine
Iowa City, Iowa 52242

I. INTRODUCTION

L-Carnitine is present in a variety of foods, but is primarily abundant in red meats, poultry, fish and dairy products. Plant-derived foods contain very little carnitine. In mammals, carnitine is synthesized from the essential amino acids L-lysine and L-methionine, via the protein-bound intermediate, ε-N-tri-methyl-L-lysine. Rates of endogenous synthesis of carnitine are sufficient to meet the normal needs of adult humans. However, it has been suggested that carnitine is a conditionally-essential nutrient for infants and perhaps growing children.

II. BACKGROUND

A. ESTIMATES OF CARNITINE ABSORPTION FROM KINETIC AND PHARMACOKINETIC STUDIES

For many years it was assumed that dietary L-carnitine was virtually totally absorbed, because very little of this amino acid normally appears in feces. Kinetic experiments in rats[1,2], dogs[3], and humans[4] all predicted a greater rate of urinary carnitine excretion than was actually observed, suggesting that not all dietary carnitine was absorbed. Indeed, balance data in humans revealed that in some individuals, the rate of carnitine intake actually exceeded its rate of excretion in urine. Results from more recent pharmacokinetic studies in humans indicated that carnitine in pharmacologic amounts given orally was less than 20% available[5,6].

B. DIRECT MEASUREMENTS OF CARNITINE TRANSPORT AND ABSORPTION IN EXPERIMENTAL ANIMALS

Carnitine transport and carnitine absorption have been studied in a variety of intestinal preparations from rats, pigs, guinea pigs and humans. Gross and Henderson[7] and Gudjonsson[8,9] demonstrated in rats in vivo that carnitine is rapidly taken up by the small intestinal mucosa, but is only slowly

[*] Supported by awards form the National Institute of Diabetes and Digestive and Kidney Diseases (DK35106) and The General Clinical Research Centers Branch, Center for Research Resources (RR00059), U.S. Public Health Service.

released into the circulation. Moreover, within the intestinal mucosa, up to 50% of the carnitine taken up from the lumen was acetylated. L-Carnitine was taken up about twice as rapidly as the D isomer, and the process was saturable, suggesting that a specific carrier for carnitine was present in the mucosal membrane[7]. Shaw et al.[10] observed active transport of L-carnitine into everted rat duodenal and jejunal sacs and rings. On the other hand, in brush border membrane vesicles prepared from young adult male rats, Li et al.[11] found no evidence of active transport of carnitine, and concluded that no carnitine transporter, per se, exists in that membrane. Gross et al.[12] concluded that carnitine is taken up by isolated guinea pig enterocytes by a facilitated diffusion process, whereas Hamilton et al.[13] reported active transport of carnitine into human intestinal biopsy specimens. More recently, Li et al.[14] have demonstrated up-regulation of carnitine transport into jejunal segments from newborn pigs, compared to suckling and weanling animals. Their work suggests that the greater rate of carnitine transport in newborn pig intestinal mucosa may be mediated by glucagon and insulin secretion (increased glucagon to insulin ratio).

C. ESTIMATES OF DIETARY CARNITINE DEGRADATION IN RATS

In the kinetic studies cited above, compartmental analysis was facilitated by injection, intravenously or intraperitoneally, of either [*methyl*-^{3}H]- or [*methyl*-^{14}C]L-carnitine. Examination of tissues, urine and, in some studies, feces, revealed no significant amounts of radiolabeled degradation products of carnitine. Subsequently, a study was undertaken specifically to examine the fate of dietary carnitine in rats[15]. [*methyl*-^{14}C]L-Carnitine (185 kBq; 86 nmol or 124 µmol) was administered orally to rats weighing 95 to 191 g. Urine and feces collected over 48 h after administration of the radioactive test dose, and serum and tissues collected at the end of 48 h were analyzed for radioactive carnitine and metabolites. Rats excreted 3.6% and 11.4% of the tracer (86 nmol) dose and the pharmacological (124 µmol) dose, respectively, in urine as [^{14}C]trimethylamine oxide. Likewise, 2.9% and 23.1% of the respective doses were found in feces as [^{14}C]γ-butyrobetaine. Virtually no radioactive metabolites were found in urine and feces when the radiolabeled carnitine was administered intravenously, or when the pharmacological dose was given to germ-free rats orally. The conclusion from this study was that carnitine was not degraded systemically, but was metabolized primarily to trimethylamine and γ-butyrobetaine by indigenous flora of the gastrointestinal tract. These results were subsequently confirmed independently[16].

III. DIETARY CARNITINE ABSORPTION AND DEGRADATION IN HUMANS

A. RATIONALE

In view of the results of kinetic and pharmacokinetic studies and investigation of dietary carnitine absorption and metabolism in rats cited above, it was reasonable to assume that dietary carnitine is not completely absorbed intact in humans. To verify this assumption, studies in humans were initiated to examine the metabolic fate of carnitine at three different dietary levels: a low-carnitine diet, a high-carnitine diet (but both within the normal range of human consumption); and a high-carnitine diet plus a pharmacological carnitine supplement. The objectives of these studies were (1) to quantify the extent of carnitine degradation in the gastrointestinal tract, and (2) to assess the rate of carnitine absorption (appearance in the circulation) in humans. These studies have been described in greater detail elsewhere[17,18].

B. PROTOCOL

Seventeen adult male volunteers were divided into three groups. Characteristics of the participants are listed in Table 1. Subjects in Group 1 (n = 6) were assigned to a diet low in carnitine (1.5 to 2.2 μmol·kg body wt^{-1}·d^{-1}). Participants in Groups 2 (n = 6) and 3 (n = 5) were provided a high-carnitine diet (8.0 to 11.9 μmol·kg body wt^{-1}·d^{-1}). In addition, subjects in Group 3 received a carnitine supplement (10.6 mmol·d^{-1}; in capsule form, consumed in equal amounts with three meals each day). For subjects in Group 3, the supplement was begun 14 days prior to the start of the study, and continued until the conclusion of the study. After 4 days of equilibration to the respective diets, each subject was given [*methyl*-^{3}H]L-carnitine (3.7 x 10^5 Bq·kg body wt^{-1}; 180-370 Bq/pmol) and 25 g of D-xylose, both dissolved in a fruit drink, with their normal morning meal on Day 1 of the study. Blood was sampled at frequent intervals for 48 h through an indwelling venous catheter, and twice daily thereafter by venipuncture, until the conclusion of the study (Day 11). Complete 24-h urine collections were obtained from 4 days prior to administration of the test doses until the conclusion of the study. Feces were collected for five days following administration of the test compounds.

Serum, urine and feces were analyzed for carnitine and radioactive species by standard techniques[17]. D-Xylose was quantified in serum (0 to 5 h) and urine (first 24-h collection)[19]. Mass spectra of major urinary and fecal metabolites were provided by the University of Iowa High Resolution Mass Spectrometry Facility.

TABLE 1
Participant Data[a]

	Low-carnitine diet[b]		High-carnitine diet[b]		High-carnitine diet plus supplement[c]	
Age (y)	27.3	(7.6)	26.2	(7.0)	28.2	(7.2)
Weight (kg)	79.1	(12.9)	82.1	(10.3)	86.1	(14.7)
Height (cm)	180	(9.0)	181	(2.9)	182	(5.0)
Serum free carn-itine[d] (μmol/L)	43.0	(5.9)	48.9	(5.4)	65.3	(9.6)
Serum total carn-itine[d] (μmol/L)	50.3	(6.2)	56.2	(5.4)	77.8	(12.4)
Creatinine excre-tion (mmol/day)	14.7	(2.7)	15.0	(1.9)	15.7	(0.81)

[a]Data are expressed as mean (standard deviation).
[b]n = 6   [c]n = 5
[d]Mean (standard deviation) of means for 12 days for each subject

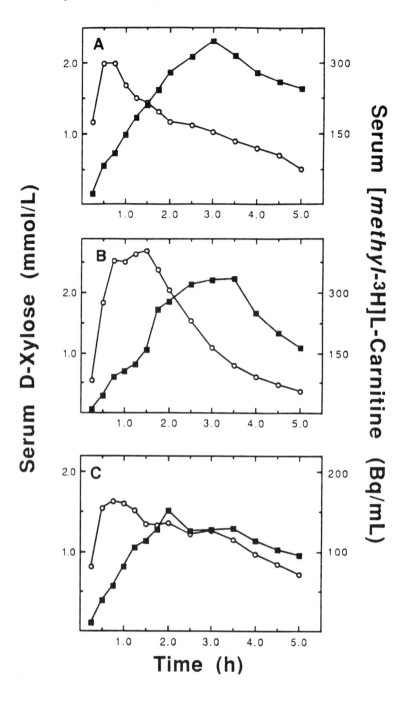

Figure 1. Serum D-xylose and [*methyl*-3H]L-carnitine concentrations following ingestion of test doses with a meal. Panel A, subject 2 (low-carnitine diet); panel B, subject 10 (high-carnitine diet); panel C, subject 17 (high-carnitine diet plus carnitine supplement). Open circles, D-xylose; closed squares, [*methyl*-3H]L-carnitine.

C. RESULTS

Absorptive function was assessed by appearance of D-xylose in serum and urine. For all subjects, D-xylose concentration in serum was highest 0.5 to 2.5 h (mean, 1.0 h) after ingestion of the test dose (Figure 1). Recovery of the test dose in urine was 23 to 41% (mean, 28%) within 24 h of oral administration. These results indicated that all subjects participating in the study had normal intestinal absorptive function[20]. [*methyl*-³H]L-Carnitine appeared in serum within 15 min after oral administration; however, the peak did not occur until 2 to 5 h (mean, 3.3 h) after ingestion of the test dose (Figure 1). Thus, carnitine was absorbed somewhat more slowly than D-xylose. Carnitine concentration in serum was not noticably affected by ingestion of a low-carnitine or high-carnitine meal, but was significantly increased 3.5 to 5 hours following ingestion of the carnitine supplement with a meal (Figure 2).

Individuals consuming the low-carnitine diet (Group 1) excreted more carnitine than they ingested (mean, 217%), whereas subjects consuming the high-carnitine diet (Group 2) and the high-carnitine diet plus supplement (Group 3) excreted less carnitine than they consumed (69% and 18%, respectively; Table 2). Of the total carnitine excreted, more than 95% was found in urine of all subjects. For 15 of 17 subjects, less than 1% of total carnitine excreted was found in feces.

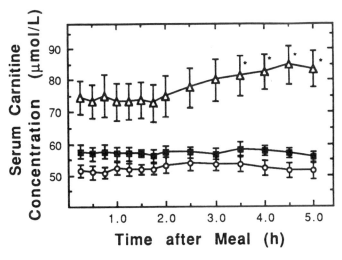

Figure 2. Effect of dietary carnitine at 3 different levels on serum carnitine concentration. Symbols indicate mean values and vertical bars indicate standard error of the mean. Group 1 (low-carnitine diet), open circles (n = 6); Group 2 (high-carnitine diet), closed squares (n = 6); and Group 3 (high-carnitine diet plus supplement), open triangles (n = 5). Vertical bars indicate standard error of the mean. Asterisks indicate value is significantly different (P < 0.05) than corresponding value at 15 min.

TABLE 2

Intake and Excretion of Carnitine by Human Subjects[a]

	Carnitine intake	Urinary carnitine excretion[b]		Fecal carnitine excretion[c]	Total carnitine excretion
	μmol/kg·day	*μmol/kg·day*	*mmol/mol creatinine*	*μmol/kg·day*	*μmol/kg·day*
Low-carnitine Diet[d]	1.79 (0.261)*	3.81 (1.30)*	20.5 (7.5)*	0.0332 (0.0576)*	3.84 (1.28)*
High-carnitine Diet[d]	9.89 (1.26)†	6.71 (1.73)†	36.7 (9.2)*	0.0332 (0.0180)*	6.74 (1.74)†
High-carnitine Diet Plus Supplement[e]	141 (22.5)‡	26.1 (4.61)‡	144 (14.2)†	0.214 (0.166)†	26.4 (4.62)‡

[a]Data are expressed as mean (standard deviation). Differences were determined by analysis of variance and Tukey's Studentized Range and Least-Square Means tests[21]. In each column, values with different superscript symbols are significantly different (P < 0.01).
[b]Mean (standard deviation) of mean values for each subject for 11 days
[c]Mean (standard deviation) of mean values for each subject for 5 days
[d]n = 6 [e]n = 5

TABLE 3

Excretion of Radiolabeled Carnitine and Metabolites by Human Subjects Following Ingestion of [methyl-3H]L-Carnitine[a]

	Urine			Feces			Total excretion as metabolites
	Carnitine	Trimethyl-amine oxide	Other metabolites	Carnitine	γ-Butyro-betaine	Other metabolites	
	(Percent of dose administered)						
Low-carnitine diet[b]	4.00 (1.42)*	16.8 (7.28)	3.24 (0.82)	0.616 (0.857)	2.68 (2.75)*	2.63 (1.33)	25.3 (7.9)*
High-carnitine diet[b]	5.14 (0.90)*†	27.7 (8.12)	4.51 (2.97)	0.319 (0.149)	1.43 (1.45)*	3.04 (1.17)	36.7 (7.2)†
High-carnitine diet plus supplement[c]	6.30 (1.36)†	31.0 (21.0)	2.85 (0.30)	0.136 (0.071)	20.5 (20.3)†	1.16 (0.82)	55.3 (3.8)‡

[a]Cummulative excretion of radiolabeled carnitine and metabolites for five days following administration of [methyl-3H]L-carnitine. Differences were determined by analysis of variance and Tukey's Studentized Range and Least-Squares Means tests[21]. In each column, values with different superscript symbols are significantly different (P < 0.02).
[b]n = 6 [c]n = 5

Chromatographic analysis of urine collected following oral administration of [*methyl*-³H]L-carnitine revealed two major radioactive species. One was coincident with carnitine, and the other was identified by chromatographic comparisons and mass spectrometry as [³H]trimethylamine oxide. Several other peaks of radioactivity were present also in urine, but they were inconsistent and quantitatively less significant. For subjects in Groups 1, 2 and 3, respectively, excretion of [³H]trimethylamine oxide accounted for 17, 28 and 31% of the administered dose of radiolabeled carnitine (Table 3). There was considerable within group variability in the extent of [³H]trimethylamine excretion, particularly among subjects in Group 3. Three participants in this group excreted relatively large amounts of this metabolite (49, 42 and 47% of the dose), whereas the remaining two subjects in Group 3 excreted only 8% of the administered dose as [³H]trimethylamine oxide. Urinary excretion of [³H]trimethylamine oxide was greatest on the first day following administration of [*methyl*-³H]L-carnitine, rapidly declined over the next four days, but was still detectable in urine by 11 days (Figure 3).

Two major radioactive compounds also were observed in feces following administration of the test dose. One was [*methyl*-³H]L-carnitine, and the other was identified by chromatographic comparisons and mass spectrometry as [³H]γ-butyrobetaine. The latter compound accounted for 3, 1 and 20% of the administered radioactive dose for subjects in Groups 1, 2 and 3, respectively (Table 3). Considerable variation between subjects in each group was noted. For subjects in Group 3, two excreted relatively large amounts of this metabolite (40 and 45% of the dose), whereas the remaining three subjects excreted relatively less (5, 12 and 0.4%). Those subjects in Group 3 that excreted large amounts of [³H]trimethylamine oxide in urine excreted relatively smaller amounts of [³H]γ-butyrobetaine in feces, and vice versa. Radioactive species other than [³H]carnitine and [³H]γ-butyrobetaine were observed in feces, but there presence was inconsistent and quantitatively less significant. Radioactive species appeared in feces predominantly within 48 hours after administration of the test dose, and were virtually undetectable within 4 to 5 days (Figure 3).

For subjects in Groups 1, 2 and 3 respectively, excretion (urinary and fecal) of all detected metabolites accounted for 25, 37 and 55% of [*methyl*-³H]-L-carnitine administered (Table 3). These results were consistent for subjects within each dietary group.

D. DISCUSSION AND CONCLUSIONS

The rapid appearance (within 15 min) of [*methyl*-³H]L-carnitine in the circulation, coupled with the relatively slow appearance of the peak of radiolabeled carnitine (Figure 1), is consistent with previous findings in rats of rapid uptake of carnitine into the intestinal mucosa and slow release into the circulation[7-9]. The concentration of carnitine in serum remains stable following ingestion of a meal containing normal amounts of carnitine (Figure 2). Sequestration of dietary carnitine in the intestinal mucosa may serve to dampen any increase in circulating carnitine concentrations, thus preventing loss of carnitine in urine. It is well known that increased circulating carnitine

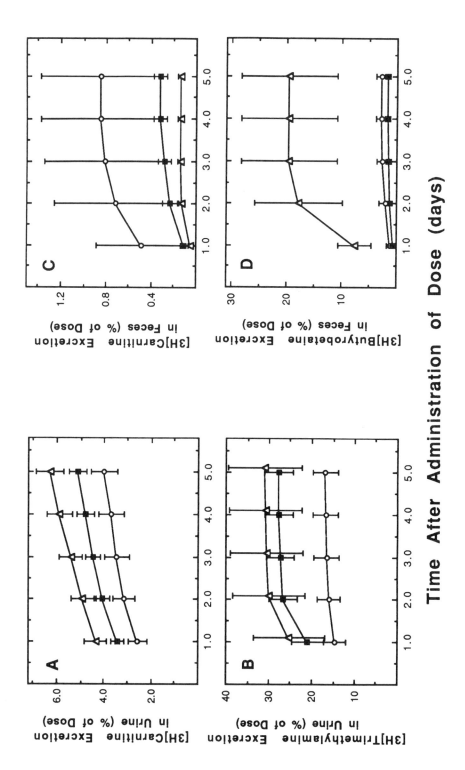

Figure 3. Cummulative excretion of radiolabeled carnitine in urine (Panel A) and feces (Panel C), radiolabeled tri-methylamine oxide in urine (Panel B) and radiolabeled γ-butyrobetaine in feces (Panel D) of humans after oral administration of [*methyl*-³H]L-carnitine Open circles, low-carnitine diet; closed squares, high-carnitine diet; open triangles, high-carnitine diet plus carnitine supplement. Vertical bars indicate standard deviation of the mean.

concentrations lead to decreased efficiency (percent of filtered load) of carnitine reabsorption by the kidney, with an attendant increase in the rate of carnitine excretion[22]. Thus, sequestration of carnitine by the intestinal mucosa may act to conserve dietary carnitine. On the other hand, saturation effects at the level of the mucosal cell due to slow release of carnitine into the circulation may limit the extent of removal of carnitine from the intestinal lumen.

Results of carnitine intake and urinary excretion data (Table 2) are consistent with previous findings from kinetic and pharmacokinetic studies that carnitine intake often exceeds carnitine excretion, particularly at high levels of carnitine intake. This result is consistent with intestinal degradation of carnitine. As the studies reviewed herein show, dietary carnitine is indeed degraded to a significant extent in humans. In a previous study it was demonstrated that radiolabeled carnitine administered intravenously was not degraded to any extent in humans[4]. Studies with conventional and germ-free rats demonstrated conclusively that intestinal flora were responsible for degradation of dietary carnitine in that species. Thus, it is reasonable to conclude that degradation of dietary carnitine in humans occurs via the action of indigenous bacterial flora in the gastrointestinal tract. The extent of degradation of dietary carnitine depends on the level of carnitine in the diet.

The extent of degradation of dietary carnitine observed in this study is likely to be underestimated, at least for subjects consuming the high-carnitine diet plus supplement. Quantification of all radioactive metabolites of [*methyl-*^{3}H]L-carnitine suggested that 55% of the dose was degraded, whereas excretion of total carnitine was only 18% of carnitine intake. Thus, 27% of carnitine intake was not accounted, nor was any endogenously-synthesized carnitine. It is possible that volatile metabolites were produced from the administered dose. In an earlier study[15], we observed no significant excretion of $^{14}CO_2$ or [^{14}C]trimethylamine by rats administered [*methyl-*^{14}C]L-carnitine orally or intravenously. However, neither those experiments in rats nor the present studies in humans were designed to quantify neutral, volatile fermentation products (for example, methane), which may account for the difference in intake versus excretion of carnitine and its observed metabolites.

IV. REFERENCES

1. Brooks, D.E., and McIntosh, J.E.A, Turnover of carnitine by rat tissues, Biochem. J., 148, 439, 1975.
2. Cederblad, G., and Lindstedt, S., Metabolism of labeled carnitine in the rat, Arch. Biochem. Biophys., 175, 173, 1976.
3. Rebouche, C.J., and Engel, A.G, Kinetic compartmental analysis of carnitine metabolism in the dog, Arch. Biochem. Biophys., 220, 60, 1983.
4. Rebouche, C.J., and Engel, A.G., Kinetic compartmental analysis of carnitine metabolism in the human carnitine deficiency syndromes. Evidence for alterations in tissue carnitine transport, J. Clin. Invest., 73, 857, 1984.

5. Segre, G., Bianchi, E., Corsi, M., D'Iddio, S., Ghiradi, O., and Maccari, F., Plasma and urine pharmacokinetics of free and of short-chain carnitine after administration of carnitine in man, Arzneim.-Forsh./Drug Res., 38, 1830, 1988.
6. Harper, P., Elwin, C.-E., and Cederblad, G., Pharmacokinetics of intravenous and oral bolus doses of L-carnitine in healthy subjects, Eur. J. Clin. Pharmacol., 35, 555, 1988.
7. Gross, C.J., and Henderson, L.M., Absorption of D- and L-carnitine by the intestine and kidney tubule in the rat, Biochim. Biophys. Acta, 772, 209, 1984.
8. Gudjonsson, H., Li, B U.K., Shug, A.L., and Olsen, W.A., Studies of carnitine metabolism in relation to intestinal absorption, Am. J. Physiol., 248, G313, 1985.
9. Gudjonsson, H., Li, B U.K., Shug, A.L., and Olsen, W.A., In vivo studies of intestinal carnitine absorption in rats, Gastroenterology, 88, 1880, 1985.
10. Shaw, R.D., Li, B U.K., Hamilton, J.W., Shug, A.L., and Olsen, W.A., Carnitine transport in rat small intestine. Am. J. Physiol. 245, G376, 1983.
11. Li, B U.K., Bummer, P.M., Hamilton, J.W., Gudjonsson, H., Zografi, G., and Olsen, W.A., Uptake of L-carnitine by rat jejunal brush border microvillous membrane vesicles. Evidence of passive diffusion, Dig. Dis. Sci., 35, 333, 1990.
12. Gross, C.J., Henderson, L.M., and Savaiano, D.A., Uptake of L-carnitine, D-carnitine and acetyl-L-carnitine by isolated guinea-pig enterocytes, Biochim. Biophys. Acta, 886, 425, 1986.
13. Hamilton, J.W., Li, B U.K., Shug, A.L., and Olsen, W.A., Carnitine transport in human intestinal biopsy specimens. Demonstration of an active transport system, Gastroenterology, 91, 10, 1986.
14. Li, B U.K., Murray, R.D., Heitlinger, L.A., McClung, H.J., Hughes, A.M., and O'Dorisio, T.M., Perinatal upregulation of intestinal transport of carnitine (C) in newborn pigs, FASEB J., 4, A653, 1990.
15. Rebouche, C.J., Mack, D.L., and Edmonson, P.F., L-Carnitine dissimilation in the gastrointestinal tract of the rat. Biochemistry, 23, 6422, 1984.
16. Seim, H., Schulze, J., and Strack, E., Catabolic pathways for high-dosed L(-)-or D(+)-carnitine in germ-free rats? Biol. Chem. Hoppe-Seyler, 366,1017, 1985.
17. Rebouche, C. J., and Chenard, C.A., Metabolic fate of dietary carnitine in human adults: identification and quantifiction of urinary and fecal metabolites, J. Nutr., 121, 539, 1991.
18. Rebouche, C.J., Quantitative estimation of absorption and degradation of a carnitine supplement by human adults, Metabolism, in press.
19. Eberts, T.J., Sample, R.H.B., Glick, M.R., and Ellis, G.H., A simplified, colorimetric micromethod for xylose in serum or urine, with phloroglucinol, Clin. Chem., 25, 1440, 1979.
20. Breiter, H.C., Craig, R.M., Levee, G., and Atkinson, A.J., Jr., Use of kinetic methods to evaluate D-xylose malabsorption in patients, J. Lab. Clin. Med., 112, 533, 1988.
21. Anonymous, *SAS User's Guide: Statistics*, 5th ed., SAS Institute, Cary, NC, 1985.

22. Engel, A.G., Rebouche, C.J., Wilson, D.M., Glasgow, A.M., Romshe, C.A., and Cruse, R.P., Primary systemic carnitine deficiency. II. Renal handling of carnitine, Neurology, 31, 819, 1981.

DISCUSSION OF THE PAPER

C. HOPPEL *(Cleveland Hts., OH)*: Does the administration of the labeled carnitine with food delay the absorption of carnitine with the resultant peak at 35 hours?

C. REBOUCHE: I do not know the answer to that question, because the tracer was given only with a meal, and not alone. However, I reiterate the observation that peak appearance of D-xylose in the circulation occurred at one hour after consumption of the test dose, also with the meal. This result does not differ from results obtained with the standard D-xylose load test for clinical evaluation of absorptive function in humans. In this text, D-xylose is not given with a meal.

C. HOPPEL *(Cleveland Hts., OH)*: If carnitine is converted to butyrobetaine in the gut, what is the influence of butyrobetaine absorption and conversion to carnitine on the plasma carnitine concentration?

C. REBOUCHE: We assume that γ-butyrobetaine appearing in the feces arises from microbial transformation of carnitine in the large intestine. I do not know if γ-butyrobetaine is absorbed in the large intestine. Conceivably, this precursor formed in the large intestine could contribute to the circulating γ-butyrobetaine pool and be hydroxylated in the liver, kidney, brain or testis. The extent to which this occurs is probably limited by the amount of γ-butyrobetaine absorbed.

E. BRASS *(Cleveland, OH)*: Professor Siliprandi and colleagues have demonstrated the potential for plasma membrane carnitine-carnitine exchange. Thus, the differential recovery of ^{3}H and carnitine "mass" could be explained by ^{3}H-carnitine entering tissue in exchange for unlabelled carnitine. This would dilute the specific activity and yield the observations you made. This phenomena has been demonstrated in animals.

C. REBOUCHE: For individuals consuming the high-carnitine diet plus carnitine supplement, only 19% of the carnitine intake is accounted for by excretion of carnitine in urine and feces. It is assumed that the remaining 81% is metabolized to other compounds. However, only 55% of the radioactive tracer was found as metabolites in urine and feces, yielding a discrepancy of 26% of the dose which is unaccounted. Carnitine-carnitine exchange between tissues and circulation cannot account for this difference, because presumably, the discrepant 26% was never absorbed intact (inferred from unlabeled carnitine balance). Moreover, because the study was conducted for 11 days, tissue and circulating pools of carnitine and radiolabeled carnitine would equilibrate (this probably occurs within 48 to 72 hours after consumption of the test dose), thus eliminating the possibility of differential excretion of tracer and tracee in the urine.

G. HUG *(Cincinnati, OH)*: What is the tissue half life in various human organ systems (liver, heart, muscle)? Is it the same or different?

C. REBOUCHE: Half-life of carnitine in tissues of mammals has not been determined; however, turnover times in various tissues of rats, dogs and humans has been estimated from kinetic studies.   In rats, turnover times for carnitine range from 0.415 hours in kidney to 105 hours in skeletal muscle[1].   In dogs the turnover time for the kinetic pool representing liver, kidney and other tissues, 7.9 hours.   Likewise for humans, these values were 191 and 11.16 hours.

1.   Brooks, D.E. and Mcintosh, J.E.A.   Turnover of carnitine by rat tissues, <u>Biochem. J.</u>, 148, 439, 1975.

ABSTRACTS TO PART I

CARNITINE DEFICIENCY IN LOWE SYNDROME
T.W. Kurczynski*, J. Marquart*, M.M. DeBeukelaer*,
C.L. Hoppel**, *Medical College of Ohio, **Case
Western Reserve University
Carnitine deficiency was documented in an infant boy
with Lowe syndrome. Studies prior to therapy included
plasma carnitines (umol/l): total 19.0, free 13.9,
short chain acyl 3.1, long chain acyl 2.0; muscle
carnitines (umol/g): total 1.25, free 0.82, short
chain acyl 0.41, long chain acyl 0.02; urine carnitine
fractions were all elevated.
Treatment with L-carnitine 500 mg. bid initially increased all plasma carnitine fractions but the pro-
gression of a renal Fanconi syndrome resulted in a
diminished response. Plasma short chain acyl
carnitine however became significantly elevated in
spite of urinary losses. These data indicate that
carnitine deficiency is an early manifestation of Lowe
syndrome and worsens as renal tubular function
deteriorates; but the effect on plasma short chain
acyl carnitine is difficult to explain solely from
renal dysfunction.

**STUDIES ON THE FORMATION AND INHIBITORY PROPERTIES OF VAL-
PROYL-, PIVALOYL-, GLUTARYL-, AND 3-METHYLGLUTARYLCARNITINE IN
CARNITINE ACYLTRANSFERASES.** G. Dai, C. Chung and L.L. Bieber, Dept.
of Biochemistry, Michigan State University, East Lansing, MI 48824.

The capacity of carnitine acetyltransferases (CAT) from rat heart
mitochondria, rat liver peroxisomes, rat liver mitochondria and pigeon breast
muscle, as well as carnitine octanoyltransferase from rat liver microsomes and
peroxisomes, and carnitine palmitoyltransferase from rat heart mitochondria, to
use unusual acylcarnitines as substrates has been determined. None of the rat
carnitine acyltransferases exchanged carnitine into glutaryl-, 3-methylglutaryl- or
valproyl-carnitine; however, these compounds are inhibitory to CAT, especially
when acetylcarnitine is the substrate. α-Methylbutyrylcarnitine is an excellent
substrate for liver CATs, but a poor substrate for heart mitochondrial CAT. The
use of a second acylcarnitine substrate in the exchange reaction enables
determination of whether the acylcarnitines bind to CAT to form an E •
acylcarnitine complex, or does not bind to a specific CAT, or whether it
undergoes catalysis. Related studies have shown that all three of the rat CATs
exhibit different kinetics with acetyl-CoA as substrate compared to other acyl-
CoAs. The E-amino-modified citraconic acid completely inhibits CAT, indicating
a role for the E-amino of lysine in binding and/or catalysis. (Supported in part
by NIH grant DK18427)

HIGH PERFORMANCE LIQUID CHROMATOGRAPHY (HPLC) OF FREE AND ACYLATED CARNITINE. M. Novak Dept. of Pediatrics, University of Miami, P.O. Box 016820, Miami, Florida 33101.

We have tested a HPLC method for the quantitative determination of free carnitine and individual acylcarnitines. The method is based on a pre-column derivation of monocarboxylates (RCOOH) using 9-anthryldiazomethane[1]

R=acyl residues with carbon atom chain length C_2-C_{24}

Simultaneously, we adopted a simple technique for the preparation of the derivatization reagent. It can be performed in most clinical laboratories not equipped to provide complicated organic synthesis.

1) 9-anthraldehyde
2) 9-anthraldehyde hydrazone
3) 9-anthryldiazomethane

According to our experience, the method is relatively simple and requires only basic HPLC equipment, measurement of the absorbance at 250 nm or fluorometric detection. It may be adapted to biological material using commercially available cartridges and an internal standard (such as heptanoylcarnitine). Quarternary amines includling betaine or γ-butyrobetaine are separated.

The high sensitivity and simplicity of this method may allow replacement in part of the widely used radioisotopic determinations and is an easy way to study individuals with abnormalities in carnitine metabolism.

Ref.: 1) Tsiguchika Yoshida et al.: J. Chromatogr., 466:175-182 (1988)

CARNITINE VALUES DURING TREATMENT WITH VALPROATE RELATIVE TO AGE OF PATIENTS AND DURATION OF THERAPY.
G.Opala*, S.Winter, H.Vance*, C.Vance, T.Hutchison, L.Linn*, T.O'Hara*, S.Szabo, A.Szabo*.
Valley Children's Hospital and UCSF, Fresno, CA. *Metabolic Research & Analysis, Inc., Fresno, CA.

Plasma total, free and acyl carnitine levels were determined in age groups (<2, 2-10, >10 years) of pediatric patients treated with valproate (VPA) as monotherapy (VPAm) (n=43) or VPA plus other antiepileptic drugs as polytherapy (VPAp) (n=92) and normals (n=89). From groups VPAm, VPAp we selected patients (n=84) with more than one carnitine test and compared carnitine levels in different periods of therapy (>2, 2-12, >12 mos). The mean free carnitine level in VPAp patients 2-10 years old was 19.5 μM/L compared to patients >10 years old with mean free levels of 26.8 μM/L (p<.05). There was no significant association between VPA therapy duration and mean plasma free carnitine levels.

Age was a risk factor for the VPAp group but not for VPAm. Duration of therapy was not a risk factor in either group.

METABOLIC EFFECTS OF PIVALATE IN ISOLATED RAT HEPATOCYTES. L.J. Ruff and E.P. Brass. Dept. of Medicine, Case Western Reserve University, Cleveland, OH 44106.

Pivalate (trimethylacetic acid) administration in man or rat has been reported to cause increased urinary carnitine loss secondary to pivaloylcarnitine generation and subsequent urinary excretion. Hepatocytes isolated from fed rats were used to study the effects of pivalate on cellular coenzyme A (CoASH) and acyl-CoA contents, and oxidative metabolism. HPLC analysis of CoASH and acyl-CoA contents in hepatocytes incubated with pivalate showed that CoASH was sequestered, as pivaloyl-CoA was formed and accumulated (mean ± SEM, *p<0.05 vs. absense of pivalate):

	CoA CONTENT IN HEPATOCYTES (nmol/10^6 cells) (N=3)			
Pivalate	CoA	AcetylCoA	SuccinylCoA	PivaloylCoA
0 mM	0.97±0.14	0.57±0.14	0.40±0.08	<0.01
0.5 mM	0.18±0.18	0.29±0.03	0.06±0.06	1.20±0.25*
1.0 mM	<0.01*	0.25±0.01	0.05±0.05*	1.25±0.13*

Enzymatic recycling analysis of the acid insoluble CoA pool revealed no pivalate-induced increase in long-chain acyl-CoA content. Pivalate (5 mM) inhibited $^{14}CO_2$ generation from 10 mM [1-^{14}C]pyruvate by 34%, but had no effect on 0.8 mM [1-^{14}C]palmitate oxidation. In contrast, propionate (5 mM) inhibited palmitate oxidation by 30%. Pivaloyl-CoA supported acylcarnitine formation at rates 10-20% those observed with equimolar acetyl-, propionyl-, isobutyryl- or isovaleryl-CoA. Hepatocyte carnitine acyltransferase activities exhibited a rank order for branched chain acyl-CoAs of isobutyryl-CoA > isovaleryl-CoA > pivaloyl-CoA. In summary, the sequestration of CoASH by pivalate was associated with the inhibition of pyruvate oxidation. As with other organic carboxylic acids, activation of pivalate to the CoA thioester is an important aspect of the compound's biochemical toxicology.

PYRIMETHAMINE AND SULFADIAZINE ADMINISTRATION PRODUCES CARNITINE DEFICIENCY. G. Sekas and H.S. Paul. University of Pittsburgh School of Medicine, Pittsburgh, PA 15213.

Pyrimethamine and sulfadiazine are antibiotics used for the treatment of toxoplasmosis. We observed a marked decrease in total carnitine concentration (16.9 vs 59 ± 10 nmole/ml) and an increase in the proportion of acylcarnitine (74% vs 25 ± 4%) in the serum of a patient being treated with these drugs. After discontinuation of these drugs, the patient's carnitine levels normalized. To determine if the carnitine deficiency was related to these drugs, we performed further studies with rats. Rats were administered either water or pyrimethamine (1.1 mg/kg) and sulfadiazine (115 mg/kg) by gastric gavage for 10 days. Total carnitine and acylcarnitine levels were measured in serum and urine. As in the patient, administration of the antibiotics to rats resulted in a 40% decrease in serum total carnitine (36.2 ± 4.6 vs 58.5 ± 6.7 nmole/ml, mean ± SE of 5 rats, p<0.01). The proportion of acylcarnitine was higher in the antibiotic group than controls (54 ± 6% vs 40 ± 2%; p<0.01). The urinary excretion of total carnitine was increased in the antibiotic group as compared to controls (1558 ± 230 vs 972 ± 90 nmole/24 hr; p<0.001). The urinary excretion of acylcarnitine was higher in the antibiotic group than controls. To determine if the formation and excretion of unusual acylcarnitines was contributing to the carnitine deficiency, serum was analyzed for acylcarnitine by mass spectrometry. No unusual acylcarnitines were detected. We conclude that treatment with pyrimethamine and sulfadiazine results in decreased serum carnitine levels by increasing its urinary excretion. Patients being treated with these drugs may require monitoring for potential carnitine deficiency.

ALTERED FAT METABOLISM IN A PIVALATE-INDUCED RAT MODEL OF CARNITINE DEFICIENCY.
P.B. Bianchi & A.T. Davis. Departments of Surgery, Michigan State University and Butterworth Hospital, Grand Rapids, MI, 49503.

To investigate the effects of excessive urinary carnitine losses, sodium pivalate, 20 mM, was given to male weanling rats to induce a carnitine deficiency. Control rats received 20mM NaHCO$_3$. After two weeks rats were fasted for 24 hours and cold stressed for 4 hours prior to tissue collection in order to maximize fat oxidation. Total tissue carnitine concentrations in the pivalate group were reduced by 33-69% while liver triglycerides (TG), and plasma TG, free fatty acids (FFA), and β-hydroxybutyrate (βOHB) were significantly increased ($p<0.01$).

		Total	Carnitine		TG	TG	FFA	βOHB
	Plasma	Heart	Muscle	Liver	Plasma	Liver	Plasma	Plasma
	nmol/ml	nmol/g	nmol/g	nmol/g	mM	mg/g	mM	mM
Pivalate	7	216	184	213	1.62	90.2	2.3	5
Control	20.4	558	508	314	0.84	8.8	1.47	2.4
Pooled SE	2.1	68	62	46	0.25	18.1	0.28	0.6

The elevated ketones and FFA and decreased tissue and plasma carnitine concentrations are consistent with abnormalities reported for children with carnitine deficiency due to propionic aciduria or Fanconi Syndrome.

CARNITINE DISTRIBUTION IN PLASMA AND BLOOD CELLS IN UREMIC PATIENTS ON ACETATE VS. BICARBONATE DIALYSIS. M. Novak, G. Zilleruelo, N. Schneider, E. Monkus, J. Strauss. Dept. of Pediatrics and Medicine, University of Miami. P.O. Box 016820, Miami, FL 33101.

The concept of carnitine (C) deficiency in uremic patients undergoing hemodialysis (HD) is still controversial. C content in red blood cells (RBC) has been often used for the assessment of C status. Leukocytes and reticulocytes also may contain significant amounts of C. However, these cells cannot be completely separated from RBC by routine or density gradient centrifugation. Thus, for assessment of C content in blood, free (FC) and esterified carnitine (AC) was measured in whole blood cells. Since dialysate buffer composition appears to influence response to L-Carnitine supplementation, we compared plasma and blood cells FC and AC in 8 patients on acetate (A) dialysis and 19 patients on bicarbonate (B) dialysis. In whole blood the proportion of AC to FC was higher than in plasma. FC in blood cells on A dialysis was significantly reduced. We also studied a subgroup of 4 patients on A dialysis and 7 patients on B dialysis by measuring pre and post-dialysis FC and AC in plasma and blood cells. A significantly lower pre-dialysis FC was observed in the A group. Plasma FC and AC significantly decreased with dialysis, but no changes were observed in blood cell carnitines. Our findings suggest different metabolic requirements for C in patients during A dialysis as compared to B dialysis. No correlation was observed between C levels, hematocrit, RBC or leukocyte count.

SERUM CARNITINE REDUCTION IN CHILDREN ON ANTICONVULSANT THERAPY WITH PHENOBARBITAL, VALPROIC ACID, PHENYTOIN AND CARBAMAZEPINE. <u>C. McGraw, G. Hug, E. Landrigan, S. Bates.</u> Children's Hospital, Cincinnati, OH 45229.

The concentration of total, free, short-chain and long-chain carnitine was compared with that of anticonvulsant drugs phenobarbital, valproic acid, phenytoin and carbamazepine in 974 serum specimens from 483 children treated for convulsions. Total carnitine was reduced in 272 (28%) of all patient specimens. The mean concentration of total, free, and short-chain carnitine for each drug treatment group was lower than control (all $p < 0.001$). Every group had specimens more than 2SD below the control mean. In mono-therapy, the concentration of phenobarbital (but not of other drugs) was inversely related to that of carnitine ($p < 0.001$). In dual drug therapy with phenobarbital + valproic acid, phenobarbital was inversely correlated with carnitine ($p < 0.01$). In 34 phenobarbital treated patients with serial serum specimens for analysis, 19 (56%) had inverse relationship between phenobarbital and carnitine concentrations, 8 (24%) had persistently low carnitine despite substantial reductions in phenobarbital, and 7 (21%) had no clear relation. Whether the reduction in serum carnitine is drug-induced or exists prior to treatment will have to be studied by measuring serum carnitine concentration before the start of anticonvulsant therapy and thereafter at regular intervals.

PART II

NOVEL NEW CARNITINE RESEARCH

Novel New Areas of Carnitine Research

A. Lee Carter, Ph.D.
Medical College of Georgia
Augusta, GA 30912

In the previous section the formation and function of short and medium chain acylcarnitines were presented. In the next section the formation and function of long chain acylcarnitine will be discussed. These subjects will be only briefly reviewed in this section. This section includes new areas of research which will likely expand in the near future.

Many studies of potential human functions and uses were carried out in animal models. In the first section the potential of carnitine in food animal production will be presented. The potential for the increase in average weight gain and quality of the meat in ruminants, swine, and fish will be discussed as well as theories as to the mechanism of these actions.

In the next chapter, the relationship between ethanol metabolism and carnitine will be presented in detail. Previous studies have indicated that the administration of carnitine prolongs the effects of ethanol metabolism.

Historically, one of the best tools for studying the active sites of enzymes has been the development of analogs of substrates. This approach with other enzymes and enzyme systems has led to the development of inhibitors and activators of reactions, many of which have become important pharmacologic agents of choice to treat certain pathological conditions. In this section, the theory and development of some important analogs of the carnitine acyltransferases will be presented in detail.

In the last chapter of this section a brief overview of carnitine functions will be presented. Also, data on the interaction of carnitine with cell membranes will be presented and discussed in detail. This is an interesting new possible function of carnitine which is independent of the acyltransferases.

CARNITINE IN FOOD ANIMAL PRODUCTION

G. L. Newton and G. J. Burtle
University of Georgia, Coastal Plain Station
Tifton

I. INTRODUCTION

Farm animals, in general, have metabolic pathways similar to those found in man and laboratory animals. They are also subject to many of the same nutritional deficiencies, metabolic disturbances and stress responses. These considerations suggest that carnitine should benefit agriculturally important animals by improving performance or increasing well-being, at least under some conditions or specific situations.

II. CARNITINE AND RUMINANT METABOLISM

The major differences in metabolism between ruminants and monogastric animals concern the utilization of products of a fermentative digestive system. While the rumen allows digestion of fibrous feeds and utilization of low quality protein to a much greater extent than possible in other animals, it also dictates that the major metabolic energy source is volatile fatty acids and that the saturation of dietary fatty acids is increased prior to absorption. Considering these points it may not be surprising that carnitine concentrations in ruminant tissues are higher than those found in most other animals.

A. SPECIAL CONSIDERATIONS

Several interesting aspects of carnitine metabolism in ruminants have been reported. Acetylcarnitine concentrations increased dramatically in sheep that were fed diets which favored acetate production by the rumen, were fasted or were alloxan-diabetic. They proposed that carnitine served to relieve "acetyl pressure" on the CoA system (the reduction in free CoA resulting from high levels of acetate). This could be especially important in ruminants since they have a lower ratio of free-CoA:acetyl-CoA than rats.[1,2] The ratio of free-CoA:acetyl-CoA may be even lower in the normal lactating cow than in the fasted sheep.[3] In conjunction with this, there is a significant production of endogenous acetate in the liver of cattle and sheep (acetate production may account for up to 70% of the free fatty acids taken up by the liver). The enzyme responsible for the final step in this process, acetylcarnitine hydrolase, shows increased activity in fasting rats and lactating cows and shows a significant decrease in alloxan-diabetic sheep.[4,5] This system provides acetate to muscle and mammary tissue and also relieves "acetyl pressure" by having the net effect of converting liver mitochondrial acetyl-CoA to free-CoA plus extramitochondrial acetate.

Carnitine palmitoyltransferase is inhibited by malonyl-CoA, however, this enzyme is also inhibited by methylmalonyl-CoA (a normal intermediate of propionate metabolism) in sheep liver mitochondria but not in mitochondria from rat or guinea-pig liver.[6] Propionate in fact does inhibit fatty acid oxidation in ruminant liver.[7,8,9]

Carnitine acyltransferase is inhibited by malonyl-CoA to an approximately equal extent in rat and sheep liver when palmitoyl-CoA is the substrate. With linoleate as substrate, 22 μM-malonyl-CoA inhibited carnitine acyltranferase by 50% in rat liver. The enzyme was inhibited by greater than 90% at malonyl-CoA concentrations of 1 μM in sheep liver.[10] Therefore, linoleate may be conserved in ruminants. This may be an important consideration since unsaturated fatty acids tend to be hydrogenated in the rumen.

The carnitine content of ruminant blood, milk and muscle decreases during winter and in response to cold exposure.[11,12] This relationship has not been extended to cover the range of heat stress, however, carnitine levels may continue to increase with increasing temperature. Lactating goats exposed to thermoneutral and moderate cold temperatures had blood acid soluble carnitine levels of 4.2 and 2.8 nM/ml respectively,[12] corresponding values for normal and ketotic cows were 9 and 30 nM/ml respectively.[13] Healthy, mid-lactation cows under summer heat stress at Tifton, Georgia had blood carnitine levels of 21.3 and 31.4 nM/ml for mature and first lactation cows respectively.[14]

B. KETOSIS

By far the most popular topic of carnitine research in ruminants has been that dealing with ketonemia, an economically important problem commonly known as pregnancy toxemia in sheep and ketosis, acetonemia and ketoacidosis in cattle. The subject has been ably reviewed.[11,15] Spontaneous occurrence of the malady is generally limited to sheep during the last weeks of gestation and to cattle during the first weeks of lactation. The outward symptoms in cattle include decreased appetite, lowered milk production, an odor of acetone from the breath and urine and central nervous system depression.

Ketosis occurs when body reserves are mobilized for productive functions and it is exacerbated by restriction of energy or protein intake. The current hypothesis is that the condition is the result of the derepression of carnitine acyltransferase, which allows large, uncontrolled, amounts of fatty acids to enter the liver mitochondria where they are metabolized to acetoacetate and subsequently form ß-hydroxybutyrate, both of which accumulate. This hypothesis does not account for all of the observed effects (several of these are listed in Table 1). Additional factors which have been suggested as causative include tissue deficiencies of biotin, thiamine, choline or tyrosine/phenylalanine/ tryptophane, and/or elevated ammonia. These factors have been suggested in light of the central nervous system effects of ruminant ketosis[15].

TABLE 1.
Some of the Observed Effects of Ketosis in Ruminants

Effect	Selected references
Transient or sustained hypoglycemia.	13, 16, 17, 15
High blood free fatty acid levels.	13, 18, 19, 20, 15
High levels of blood ß-hydroxybutyrate and acetoacetate.	13, 16, 15
Carnitine increases in blood, milk, tissues and urine.	13, 18, 2, 11
Ratio of free-/acetyl- for carnitine and CoA reduced in tissues, blood and liver.	1, 11, 18, 2
Carnitine synthesis accelerated.	21, 22
Creatine synthesis restricted.	21
Carnitine acyltransferase activity increased.	8, 23, 15
Plasma gamma-globulin reduced.	20

Ruminant ketosis has been produced experimentally by feed restriction,[16] feed restriction coupled with thyroxine injections[17] and alloxan administration.[11] Spontaneously ketotic animals have been used as research subjects by several investigators as well. Control measures investigated have included administration of i.v. glucose, propionate, cortisol, amino acids, pantothenic acid and carnitine as well as oral propionate and carnitine.[17,16,3,15]

Although ruminant ketosis is usually characterized by at least a transient hypoglycemia, gluconeogenesis may not be significantly reduced[7] and gluconeogenesis may require at least a basal rate of ketogenesis from long chain fatty acids.[24,25] Although the presence of acetoacetate and ß-hydroxybutyrate <u>per</u> <u>sae</u> are not considered to be pathological,[15] it has been reported[26] that the tricarboxylic acid cycle is inhibited by acetoacetate and ß-hydroxybutyrate in the presence of carnitine.

Excess ammonia from tissue mobilization of amino acids is an appealing candidate as a causative factor.[15] The effects of ketosis on the central nervous system are somewhat similar to those of ammonia toxicosis. Excess ammonia favors the production of glutamate and glutamine from α-ketoglutarate in brain tissue, thereby limiting oxidative metabolism and apparently interfering with the replenishing of tricarboxylic acid cycle intermediates.[27] In fact, plasma glutamate was significantly increased after subjecting lactating cattle to a treatment of low energy intake plus thyroxine injections for 7 days. The treatment with the highest glutamate level (carnitine injections) had the lowest mean score for signs of ketosis, the highest mean blood glucose and mean blood ketones within the normal range.[17] Carnitine has been shown to increase the mitochondrial glutamate of ammonia challenged mice,[28] which is consistent with these observations in ketotic cattle. It is recognized that the effectiveness of carnitine as a treatment for hyperammonia is the subject of debate[29] and also, mobilization of tissue protein reserves, which is necessary for ammonia release, may be limited by fatty acids or a fat metabolite.[30]

The effects of ketosis on carnitine and CoA are dramatic. Total acid soluble carnitine concentration increases in blood, muscle and milk.[21,18,2,13] These increases are often the result of decreases in free carnitine and increases in acetylcarnitine. The daily excretion of carnitine in milk may double, even though total milk production is significantly reduced. Like carnitine, the ratio of free-CoA/acetyl-CoA is drastically reduced although free-CoA concentrations may change very little. The increased carnitine is the result of increased synthesis in the liver, although liver carnitine concentrations may change little or increase as much as 28-fold.[13,22,21] The increase in carnitine synthesis is accompanied by a decrease in creatine synthesis and any effects on choline or other methylated metabolites have apparently not been investigated. Glucocorticoid administration has been shown to increase the free/acetyl ratio of both carnitine and CoA in tissues of cattle[3].

The basis for use of carnitine as a treatment measure for ruminant ketosis is in light of the decreases often observed in free carnitine. Results have been variable with the apparent best response from oral D,L-carnitine.[16,17,19,20] Increased carnitine synthesis may be a protective response, rather than pathogenic, such that supplying exogenous carnitine prior to the onset of maximum synthesis could be protective. It is however, apparent that a sustained carnitine deficiency is not an underlying cause of ruminant ketosis.

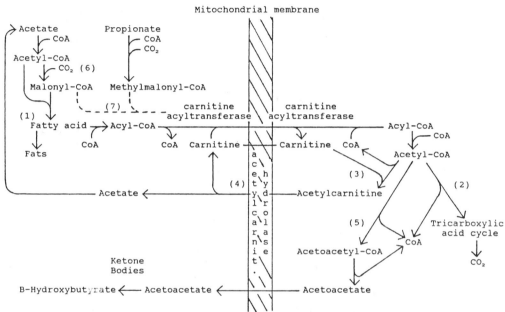

FIGURE 1. Schematic of fatty acid metabolism in ruminant liver. Events and control points involved in ruminant ketosis: 1) Free fatty acids are mobilized from tissues and begin entering liver mitochondria, 2) Liver cell energy needs are met and tricarboxylic acid cycle is inhibited, 3) Acetyl-CoA is shunted to acetylcarnitine, 4) Acetyl-carnitine hydroxylase is inhibited resulting in the accumulation of acetyl-carnitine, 5) The demand for free-CoA forces acetyl CoA to be metabolized to acetoacetate, 6) Acetyl-CoA carboxylase is inhibited by hormones or metabolites, 7) Malonyl-CoA falls, derepressing carnitine acyltransferase I, thus increasing fatty acid entrance into the mitochondria.

Interest has been shown in carnitine acyltransferase and control of its activity as a causative factor in ruminant ketosis. However, this system does not adequately account for the effects. Reports concerning fatty acid metabolism in ruminant liver may be enlightening.[5,9,4] As noted above acetate production can account for 70% of the carbon from palmitate taken up by the liver. On the other hand ß-hydroxybutyrate and acetoacetate can account for 65% of the carbon from palmitate taken up by the liver. It is not clear to what extremes these percentages apply, however it is apparent that they depend on the activity of acetylcarnitine hydrolase. The fact that the activity of this enzyme is reported to significantly decrease in alloxan-diabetic sheep adds support to its involvement. We propose that spontaneous ruminant ketosis is a result of some factor or combination of factors which derepress carnitine acyltransferase while simultaneously repressing acetylcarnitine hydrolase. This theory is presented schematically in Figure 1.

III. CARNITINE IN SWINE NUTRITION AND METABOLISM

Overall metabolism, and that of carnitine, in pigs is much closer to that of man and laboratory animals than is that of ruminants. As might be expected, there are more reports dealing with carnitine in swine as a model for humans than dealing with pork production.

A. NEONATAL PIGS

The neonatal pig has been found to be a suitable model for the study of carnitine metabolism applicable to newborn humans.[31,32,33] Carnitine levels increase in most tissues with colostrum intake and time and are dependant upon exogenous sources for maintenance of normal values.

It was found that carnitine palmitoyltransferase in neonatal pig liver mitochondria increased up to 24 hours but changed little from that point to 24 days of age, even though liver cell mitochondria from 24-day-old pigs oxidized palmitoyl-CoA, in the presence of carnitine, at a rate double that of mitochondria from 1-day-old pigs.[34] It has also been reported that i.v. carnitine does not affect palmitate oxidation in the intact animal but that pigs from sows fed 8% added fat during gestation oxidize palmitate to a significantly greater extent than pigs from sows fed no added fat.[35] In related work, it was found that pigs from sows fed added fat during gestation had increased levels of serum free carnitine while pigs from sows fed a high lysine diet had decreased serum free carnitine levels.[36] This latter observation is interesting in light of the report that high lysine intake decreases carnitine excretion in the rat[37] and suggest that the relationship holds true for carnitine transfer to the fetus.

The addition of oleate plus carnitine to hepotocytes from 48-hour-old pigs increased the rate of gluconeogenesis from lactate by 30%.[38] Carnitine addition reportedly did not influence the utilization of medium-chain triglycerides[39] while others have reported that carnitine infusion increases the rate of medium-chain fatty acid oxidation in newborn pigs.[40]

B. WEANLING PIGS

Supplementation of 4 kg mini pigs with carnitine during total parenteral nutrition has been shown to increase their energy gain from exogenous fat and to increase their nitrogen retention four fold.[41] These changes were accompanied by decreases in triglycerides and free fatty acids and increased ß-hydroxybutyrate in serum. In a series of trials during which 4-week-old pigs were fed supplemental carnitine levels ranging from 25 to 6000 mg/kg of diet it was found that the response to carnitine is curvilinear and dependant upon the lysine level of the diet.[42,43,44,45,46] Supplemental carnitine resulted in increased growth rate at moderately high lysine intakes, but tended to depress growth rate when lysine levels were near or below published requirements or in considerable excess. The relationship of carnitine and high levels of lysine to feed efficiency and the rate of body weight gain is illustrated in Figure 2. Additionally it was found that pigs which were light weight at weaning, relative to the mean of the groups, (generally those of approximately 5 kg or less) were more likely to have a positive growth response to supplemental carnitine over the 4-week nursery period than heavier pigs. This was true even in light of the fact that heavier pigs had a more positive response in feed intake and growth rate during the first 4 days after weaning. It was later determined that the gains of larger pigs, made during these first few days, could be maintained if supplemental carnitine levels were drastically reduced after 7 days.

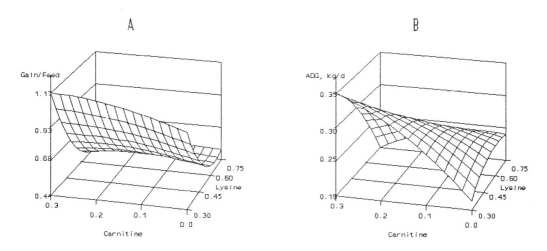

FIGURE 2.

The response in feed efficiency (A) and growth rate (B) (ADG = average daily gain) to varying intakes of L-carnitine and L-lysine (g/kg body wt/day).

The other interesting observation from our studies was tha
the response to supplemental carnitine may be sex dependant i
addition to being body weight and diet dependant, at least unde
the conditions of our trials. Female pigs, especially if they ar
lighter than the median weight, show significant improvements i
growth rate or efficiency of feed usage in response to supplementa
carnitine when compared to male castrate pigs (Figure 3). Thi
response is not female specific, as intact male pigs respond t
supplemental carnitine in a manner that is similar to that o
female pigs, and their response may be less body weight dependant

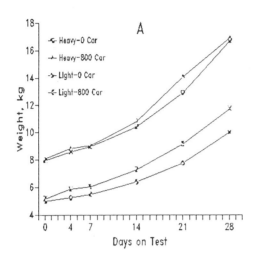
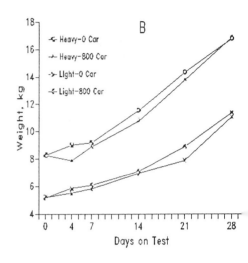

FIGURE 3. Body weight gain of heavier and lighter female (A) an
male castrate (B) pigs fed diets containing 0 or 80
mg/kg supplemental L-carnitine.

Kansas workers, using 3-week-old pigs of approximatel
5.8 kg mean weight found that the addition of carnitine (50
or 1000 mg/kg of diet for first 2 weeks reduced to 250 and 50
for an additional 3 weeks) resulted in an improvement in fee
utilization efficiency in diets containing soybean oil. I
another trial they found that supplemental carnitine improve
growth rate during the first 2 weeks post-weaning, and fee
efficiency during weeks 3 to 5 post-weaning, when pigs wer
fed diets containing 5% soybean oil. No differential effect
of carnitine due to pig sex class were observed.[47]

C. FINISHING PIGS

Research on carnitine metabolism in pigs which ar
approaching market weight is limited. Two trials have bee
reported during which 75 kg pigs were fed diets containing th
minimum requirement for lysine and either 0, 5 or 10 mg/k
supplemental carnitine.[48] When results of the trials wer

combined, there was a curvilinear response in growth rate for the first 14 days and in backfat thickness, at slaughter weight, due to carnitine intake. Growth rate and backfat were both increased at higher carnitine intakes (greater than .4 mg/kg body weight/day). There was also a differential response due to sex class with female pigs responding to a much greater degree than male castrates. Overall, the results were somewhat similar to those that would be expected if additional energy were added to a diet containing marginal lysine. The significant point of these trials was that small quantities of supplemental carnitine were biologically active in pigs that were approaching sexual maturity.

D. CONCLUSIONS CONCERNING SWINE

The carnitine status of neonatal pigs is influenced by the diet composition of their dams and neonatal pigs require an exogenous source of carnitine. The extent to which their growth and survival depends upon this supply of carnitine is incompletely understood, but situations may conceivably arise during which carnitine supplementation would be justified.

It appears that the supplementation of weanling pigs with carnitine has commercial potential. The major response may be improved growth rate when fed with lower energy diets and improved feed efficiency when fed with added fat diets. A slight excess of dietary lysine is necessary for achieving positive response. Differential responses of pigs due to their sex class may be related to their age or the type of diet fed. In addition to the conventional modes of activity associated with carnitine, pigs sometimes experience excess ammonia production from their large intestine when first weaned to vegetable protein diets. The reported effects of carnitine in lowering ammonia toxicosis could be a factor.[49] Some pigs and strains of pigs are susceptible to a condition that is analogous to halothane sensitivity in humans.[50] Such pigs should benefit from carnitine supplementation,[51] since the meat quality of these pigs is usually poor and exposing them to stress often results in death.

Optimum dietary carnitine levels would appear to be between 800 and 1200 mg/kg for the first 1 or 2 weeks post-weaning with the level dropped to 50 to 500 mg/kg after that time. The demonstrated activity in finishing pigs of low levels of carnitine supplementation suggest that additional research may show a benefit of the practice.

IV. CARNITINE IN FISH NUTRITION AND METABOLISM

Fish culture is increasing in importance, as wild stocks are declining in many areas of the world and the consumer demand for fish is increasing. An understanding of fish metabolism and nutritional requirements has therefore become essential. Differences between the metabolism of fishes and mammals appear to be related to a respiratory system adapted to function in water rather than air and the fact that fishes are poikilothermic rather than homoiothermic.

A. FATTY ACID OXIDATION IN FISH

Fish liver mitochondria are somewhat unique in that they are capable of oxidizing long-chain fatty acids in the absence of carnitine.[52] The acylcarnitine dependant system is also present in these mitochondria[53,54] and the relative degree to which the systems are utilized appears to be temperature dependant.[55] Carnitine independent fatty acid oxidation is less temperature sensitive than the carnitine dependant pathway; its importance therefore increases as temperature declines. In Lake Charr liver mitochondria incubated at 1°C, the activity of the carnitine independent pathway is reported to be approximately 60% of that of the carnitine dependant pathway while its relative activity dropped to approximately 50% at 10°C and 40% at 20°C.[55] In two species of Antarctic fish and two species of ecotypically similar temperate fish, lipid and aerobic metabolism were found to be more important in the polar fishes, but no detectable carnitine palmitoyltransferase activity was found in the white muscle of either species.[56] This suggest that carnitine may be of greatest importance in internal organ and red muscle metabolism. On the other hand white muscle from Lake Charr requires carnitine for fatty acid oxidation[55] and both carnitine independent and dependant activities have been reported in Rainbow Trout white muscle.[53]

Ketone bodies appear to serve much the same function in elasmobranchs as in mammals, but this is not true for teleosts.[57] Ketone bodies are not utilized by most tissues in these fishes and levels may depend upon the relative ketogenic and ketolytic activities of the liver. These activities are temperature dependant and appear to be related to the changing needs for acetoacetate for cholesterol synthesis (for temperature dependant membrane fluidity changes).[55]

B. CARNITINE SUPPLEMENTATION FOR FISH

Italian scientists observed that carnitine levels in blood and muscle of hatchery-reared fish were generally lower than those in wild fish of the same species.[58] Based on these observations they began a series of studies concerning the effects of supplemental carnitine on growth and lipid, nucleic acid and protein metabolism of cultured Sea Bass.[58,59,60,61] They found that Sea Bass fry in water containing L-carnitine grew significantly faster than fry not exposed to carnitine, while fry reared in water containing D-carnitine grew significantly slower and had higher mortality than control fish. Fry treated with L-carnitine contained more protein and less total lipid than unsupplemented fry. This effect was related to their more rapid utilization of retained egg lipids and a more complete transition from protein to lipid as the primary energy source, than found in nontreated fry. Later studies, during which carnitine was dissolved in water and applied to food pellets that were then dried (concentrations adjusted to supply fish with 250 mg/kg of body weight/day), confirmed the initial results and also illustrated that L-carnitine ameliorated the detrimental effects of feeding high fat diets. A study utilizing tritiated substrates confirmed increased protein deposition in carnitine treated fish but no significant changes in DNA or RNA syntheses were detected.

A Japanese patent application describes increases in weight gain or survival in fishes, eels, shellfish and prawn supplemented with .001 to 1.0% DL-carnitine chloride.[62]

The initial interest in carnitine supplementation of channel catfish diets at the University of Georgia was based on reports that carnitine reduced the effects of ammonia toxicity in mice.[63,64] Dense culture and high feeding rates often produce pond water ammonia levels which may potentially affect growth and feed utilization efficiency of channel catfish. Catfish fingerlings that were cultured in water containing .05 mg/l unionized ammonia exhibited a growth rate depression that was relieved by the inclusion of 1000 mg of supplemental L-carnitine/kg of feed.[65] Unexpectedly, fish reared in ammonia-free water also showed a growth rate response to supplemental carnitine, which led to additional investigation. Results were somewhat similar to those observed for weanling pigs in that the response was curvilinear. Supplementation of 500 mg/kg of diet produced a nonsignificant increase in growth rate, 1000 mg/kg produced a significant increase in growth rate and feed efficiency and the response to 2000 mg/kg was approximately equal to that observed with 500 mg/kg. Fish fed 2000 mg/kg, however, did have significantly less lipid in their fillets than fish fed the control diet. This reduced lipid content appeared to be mainly associated with the lateral line (red) muscle.[66] These effects and tissue carnitine concentrations are shown in Table 2.

TABLE 2.
Growth Feed Efficiency, Muscle Lipid and Tissue Carnitine
Concentrations in Channel Catfish Fingerlings Fed Carnitine
Supplemented Diets

Dietary carnitine (mg/kg)	Weight* gain %	Gain/ feed	Muscle lipid %	Muscle carnitine nM/g	Liver carnitine nM/g
0	392.5[a]	.70[a]	7.48[a]	183[a]	3.4[a]
500	426.2[ab]	.75[ab]	7.88[a]	1155[b]	13.0[b]
1000	453.3[b]	.81[b]	7.72[a]	5592[c]	14.3[b]
2000	415.2[ab]	.74[ab]	6.26[b]	12270[d]	18.5[b]

*After feeding diets for 12 weeks to fish weighing 14.7
g initially.
[abc]Means in the same column with different superscripts
are different, $P < .05$.

In response to the low muscle and liver levels of
carnitine found in channel catfish fed diets without
supplemental carnitine, the carnitine biosynthetic capacity of
catfish was evaluated.[67] The results indicated that channel
catfish liver metabolizes g-butyrobetaine to carnitine at a
rate equivalent to 3% of the rate observed in bovine liver
controls, and the biosynthetic capacity was nil or nonexistent
in other tissues of the fish. At this time, it appears that
carnitine metabolism in catfish may be somewhat similar to
that found in many neonatal mammals.

C. OUTLOOK FOR CARNITINE IN FISH CULTURE
    Mitochondrial metabolism of fatty acids in fish liver can
occur in the absence of carnitine, although the carnitine
dependant pathway is also present. Results from feeding
trials suggest that the carnitine dependant metabolism of
fatty acids is more efficient than the carnitine independent
pathway, and that it is limited by carnitine availability in
cultured fishes. Depending upon economic considerations,
carnitine may become a common ingredient in feeds for farmed
fish.

V. SUMMARY

    Although carnitine has been studied in man and under
laboratory conditions for many years, its effectiveness in
promoting the performance and well-being of our domestic
animals has only recently received significant attention. A
place for carnitine in the diets of swine and fish has been

established and continued research may find that it has a place in the production of ruminants. Although not reviewed here, the investigation of the importance of carnitine in reproductive function[68,69] may reveal aspects of economic importance. Although information is available concerning carnitine metabolism in birds and poultry, the research has primarily been in vitro with no production research that the results could be correlated with. Especially when the larger requirement for liable methyl group donors in poultry is considered,[70] carnitine in poultry production may be an exciting area for research.

VI. REFERENCES

1. Snoswell, A. M. and Henderson, G. D., Aspects of carnitine ester metabolism in sheep liver, Biochem. J., 119, 59, 1970.
2. Snoswell, A. M. and Koundakjian, P. P., Relationship between carnitine and coenzyme A esters in tissues of normal and alloxan-diabetic sheep, Biochem. J., 127, 133, 1972.
3. Baird, G. D., Heitzman, R. J., and Snoswell, A. M., Effects of a glucocorticoid on the concentrations of Co-A and carnitine esters and on redox state in bovine liver, Eur. J. Biochem., 29, 104, 1972.
4. Costa, N. D. and Snoswell, A. M., Enzymatic hydrolysis of acetylcarnitine in liver from rats, sheep and cows, Biochem. J., 152, 161, 1975.
5. Costa, N. D., McIntosh, G. H., and Snoswell A. M., Production of endogenous acetate by the liver in lactating ewes, Aust. J. Biol. Sci., 29, 33, 1976.
6. Brindle, N. P. J., Zammit, V. A., and Pogson, C. I., Regulation of carnitine palmitogltransferase activity by molonyl-Co-A in mictochondria from sheep liver, a tissue with low capacity for fatty acid synthesis, Biochem. J., 232, 177, 1985.
7. Aiello, R. J., Kenna, T. M., and Herbein, J. H., Hepatic gluconeogenic and ketogenic interrelationships in the lactating cow, J. Dairy Sci., 67, 1705, 1984.
8. Lomax, M. A., Donaldson, I. A., and Pogson, C. I., The control of fatty acid metabolism in liver cells from fed and starved sheep, Biochem. J., 214, 553, 1983.
9. Jesse, B. W., Emery, R. S., and Thomas, J. W., Control of bovine hepatic fatty acid oxidation, J. Dairy Sci. 69, 2290, 1986.
10. Reid, J. C. W. and Husbands, D. R., Oxidative metabolism of long-chain fatty acids in mitochondria from sheep and rat liver, Biochem. J. 225, 233, 1985.
11. Snoswell, A. M. and Henderson, G. D., Carnitine and metabolism in ruminant animals, in Proc. Virginia Lazenby

O-Hara <u>Biochem. Symp., Carnitine Biosynthesis, Metabolism, and Functions</u>, Frankel, R. A., and McGarry, J. D., Eds., Univ. Adelaide, Glen Osmond, Australia, 1979. Academic, New York, NY, 1980, 191.

12. Thomson, E. M., Snoswell, A. M., Clarke, P. L., and Thompson, G. E., Effect of cold exposure on mammary gland uptake of fat precursors and secretion of milk fat and carnitine in the goat, <u>Quat. J. Exper. Physiol.</u>, 64,7, 1979.

13. Erfle, J. D., Saucer, F. D., and Fisher, L. J., Interrelationships between milk carnitine and blood and milk components and tissue carnitine in normal and ketotic cows, <u>J. Dairy Sci.</u>, 57, 671, 1974.

14. Newton, G. L., Carter, A. L., and West, J. W., unpublished data, 1990.

15. Lindsay, D. B. and Oddy, V. H., Pregnancy toxaemia in sheep - a review , in <u>Proc. Symp. Recent Advances in Animal Nutrition in Australia</u>, Cumming, R. B., Ed., Univ. New England Publ. Unit, Armidale, 1985, paper no. 30.

16. Erfle, J. D., Fisher, L. J., and Saucer, F., Effect of infusion of carnitine and glucose on blood glucose, ketones, and free fatty acids of ketotic cows, <u>J. Dairy Sci.</u>, 54, 673, 1971.

17. Kellogg, D. W. and D. D. Miller, Effects of thyroxine, carnitine, and pantothenic acid plus cystein on milk production and certain metabolites of cows receiving a low-energy ration, New Mexico State Univ. Agric. Exp. Sta. Bull. 654, 1977, 23 pp.

18. Erfle, J. D., Fisher, L. J., and Sauer, F., Carnitine and acetylcarnitine in the milk of normal and ketotic cows, <u>J. Dairy Sci.</u>, 53, 486, 1970.

19. Buonaccorsi, A. and Della Croce, G., DL carnitine by oral treatment in bovine ketosis, in <u>Proc. 20th World Veterinary Congress</u>, Thessaloniki, Greece, 1975, publ. 1976, 3, 1921.

20. Ceci, L., Totaro, S., Smaldone, M., and Bilzi, A. [Nutritional factors in disease: features of preketotic acetonaemia and its therapeutic control.] Nutrizione e pathologia condizionata: aspetti dellácetonemia prechetosica e suo controllo therapeutico, <u>Clinica Veterinaria</u>, 101, 743, 1978.

21. Henderson, G. D., Xue, G.-P., and Snoswell, A. M., Carnitine and creatine content of tissues of normal and alloxan-diabetic sheep and rats, <u>Comp. Biochem. Physiol.</u>, 76B, 295, 1983.

22. Snoswell, A. M. and McIntosh, G. H., The liver as the site of carnitine biosynthesis in sheep with alloxan-induced diabetes, <u>Aust. J. Biol. Sci.</u>, 27, 645, 1974.

23. Butler, S. M., Faulkner, A., Zammit, V. A., and Vernon, R. G., Fatty acid metabolism of the perfused caudate lobe from livers of fed and fasted non-pregnant and fasted

late pregnant ewes, Comp. Biochem. Physiol., 91B, 25, 1988.

24. Chow, J. C., Planck-Meyer, C., and Jesse, B. W., Gluconeogenic dependence on ketogenesis in isolated sheep hepatocytes, J. Dairy Sci., 73, 693, 1990.

25. Aiello, R. J. and Armentano, L. E., Effects of volatile fatty acids on propionate metabolism and gluconeogenesis in caprine hepatocytes, J. Dairy Sci., 70, 2504, 1987.

26. Dynnik, V. V., Kim, Y. V., Maevskii, E. I., and Khasanov, V. A., Stoichiometric control by Co-A thioesters and inhibition by malate of the tricarboxylate cycle., Stud. Biophys., 102, 197, 1984. (Chem. Abstr. 102:127443b).

27. White, A., Handler, P., and Smith, E. L., Principles of Biochemistry, fourth edition, McGraw-Hill, New York, 1968, 863.

28. Miguez, M. P., Costell, M., O'Connor, J. E., and Grisolia, S., L-carnitine and hyerammonemia, Adv. Clin. Enzymol., 4, 111, 1986.

29. Singh, K. R., Deshmukh, G. D., and Deshmukh, D. R., Failure of carnitine to protect mice against ammonia toxicity, FASEB J., 4, A800, 1990.

30. Goodman, M. N., Lowell, B., Belur, E., and Ruderman, N. H., Sites of protein conservation and loss during starvation: Influence of adiposity, Am. J. Physiol., 246, E383, 1984.

31. Baltzell, J. K., Bazer, F. W., Miguel, S. G., and Borum, P. R., The neonatal piglet as a model for human neonatal carnitine metabolism, J. Nutr., 117, 754, 1987.

32. Coffey, M. T., Shireaan, R. B., Jones, E. E., and Herman, D. L., Effect of L. carnitine supplementation on carnitine status of neonatal pigs, FASEB J., 3, A1264, 1989.

33. Borum, P. R., Normal development of tissue carnitine concentrations in neonatal piglets, FASEB J., 5, A916, 1991.

34. Bieber, L. L., Markwell, M. A. K., Blair, M., and Helmrath, T. A., Studies on the development of carnitine polmitoyltransferase and fatty acid oxidation in liver mitochondria of neonatal pigs, Biochim. Biophys Acta., 326, 145, 1973.

35. Garber, M. J. and Froseth, J. A., Effect of intravenous carnitine administration on palmitic acid oxidation in fasted neonatal pigs, in Proc. Western Section Am. Soc. Anim. Sci., 35, 26, 1984.

36. Lewis, L. and Froseth, J. A., The effect of fat and lysine in sow gestation diets on body composition of neonatal pigs, pig and sow serum carnitine status, and pig liver succinate dehydrogenase (SDH) activity, Proc. Western Section Am. Soc. Anim. Sci., 39, 173, 1988.

37. Davis, A. T., Effect of dietary cations upon trimethyllysine and carnitine content in the rat, FASEB J., 4, A801, 1990.

38. Duee, P. H., Pegorier, J. P., Peret, J., and Girard, J., Separate effects of fatty acid oxidation and glucagon on gluconeogenesis in isolated hepatocytes from newborn pigs, <u>Biol. Neonate</u>, 47, 77, 1985.

39. Benevenga, N. J., Steinman-Goldsworthy, J. K., Crenshaw, T. D.m and Odle, J., Utilization of medium-chain triglycerides by neonatal piglets: I. Effects on milk consumption and body fuel utilization, <u>J. Anim. Sci.</u>, 67, 3331, 1989.

40. van Kempen, T. A. T. G. and Odle, J., Does supplemental carnitine affect octanoate oxidation in colostrum-deprived newborn pigs? <u>FASEB J.</u>, A593, 1991.

41. Bohles, H., Segerer, H., and Fekl, W., Improved N-retention during L-carnitine-supplemented total parenteral nutrition, <u>J. Parenteral Enteral Nutr.</u>, 8, 9, 1984.

42. Newton, G. L. and Haydon, K. D., Carnitine in nursery pig diets, Univ. of Georgia Swine Report, Univ. Georgia Coll. Ag. Spec. Publ. 44, 45, 1987.

43. Newton, G. L. and Haydon, K. D., Carnitine in nursery pig diets, <u>J. Anim. Sci.</u>, 66, Suppl. 1, 40, 1988.

44. Newton, G. L. and Haydon, K.D., Carnitine supplementation of swine diets, presented at Lonza Symp. Anim. Nutr., Visp, Switzerland, Oct. 24, 1989, section 4.

45. Newton, G. L., Haydon, K. D., and Blum, S. A., An interaction of supplemental dietary carnitine and lysine in weanling pigs. <u>J. Anim. Sci.</u>, submitted, 1991.

46. Newton, G. L., Haydon, K. D., Carter, A. L., and Blum. S. A., An influence of sex class and body weight on the response of weanling pigs to supplemental dietary carnitine, <u>J. Anim. Sci.</u>, submitted, 1991.

47. Weeden, T. L., Nelssen, J. L., Hansen, J. A., Fitzner, G. E., Goodband, R. D., Li, D. F., and Blum, S. A., Effect of L-carnitine on starter pig performance and fat utilization, Kansas State Swine Day Report, Kansas Ag. Exp. Sta. Progress Rept. 610, 56, 1990.

48. Newton, G. L. and Haydon, K. D., Carnitine supplementation for finishing pigs, <u>J. Anim. Sci.</u>, 67(Suppl. 1), 267, 1989.

49. O'Connor, J. E., Costell, M. Miguez, M. P., and Grisolia, S., Influence of the route of administration on the protective effect of L-carnitine against acute hyperammonemia, <u>Biochem. Pharmacol.</u>, 35, 3173, 1986.

50. Schaefer, A. L., Doornenbal, H., Tong, A. K. W., Murray, A. C., and Sather, A. P., Effect of time off feed on blood acid-base homeostasis in pigs differing in their reaction to halothane, <u>Can. J. Anim. Sci.</u>, 67, 427, 1987.

51. Branca, D., Toninello, A., Scutari, G., Florian, M., Siliprandi, N., Vincenti, E., and Giron, G. P., Involvement of long-chain acyl Co-A in the antagonistic effects of halothane and L-carnitine on mitochondrial

energy-linked processes, <u>Biochem. Biophys. Res. Commun.</u>, 139, 303, 1986.

52. Brown, W. D. and Tappel, A. L., Fatty acid oxidation by carp liver mitochondria, <u>Arch. Biochem. Biophys.</u>, 85, 149, 1959.

53. Bilinski, E. and Jonas, R. E. E., Effects of coenzyme A and carnitine on fatty acid oxidation by rainbow trout mitochondria, <u>J. Fish. Res. Board Canada</u>, 27, 857, 1970.

54. Henderson, R. J. and Tocher, D. R., The lipid composition and biochemistry of freshwater fish, <u>Prog. Lipid. Res.</u>, 26, 281, 1987.

55. Ballantyne, J. S., Flannigan, D., and White, T. B., Effects of temperature on the oxidation of fatty acids, acyl carnitines, and ketone bodies by mitochondria isolated from the liver of the lake charr, <u>Salvelinus namaycush</u>, <u>Can. J. Fish. Aquat. Sci.</u>, 46, 950, 1989.

56. Crockett, E. L. and Sidell, B. D., Some pathways of energy metabolism are cold adapted in Antarctic fishes, <u>Physiol. Zool.</u>, 63, 472, 1990.

57. Zammit, V. A. and Newsholme, E. A., Activities of enzymes of fat and ketone-body metabolism and effects of starvation on blood concentrations of glucose and fat fuels in teleost and elasmobranch fish, <u>Biochem. J.</u>, 184, 313, 1979.

58. Santulli, A. and D'Amelio, V., The effects of carnitine on the growth of sea bass, <u>Dicentrarchus labrax</u> L., fry, <u>J. Fish Biol.</u>, 28, 81, 1986.

59. Santulli, A. and D'Amelio, V., Effects of supplemental dietary carnitine on growth and lipid metabolism of hatchery-reared sea bass (<u>Dicentrarchus labrax</u> L.), <u>Aquaculture</u>, 59, 177, 1986.

60. Santulli, A. M., Curatolo, A., and D'Amelio, V., Carnitine administration to sea bass [<u>Dicentrarchus labrax</u> (L.)] during feeding on a fat diet: Modification of plasma lipid levels and lipoprotein pattern, <u>Aquaculture</u>, 68, 345, 1988.

61. Santulli, A., Puccia, E., and D'Amelio, V., Preliminary study on the effect of short-term carnitine treatment on nucleic acids and protein metabolism in sea bass (<u>Dicentrarchus labrax</u> L.) fry. <u>Aquaculture</u>, 87, 85, 1990.

62. Ogura, A., Feed additives for cultivated fishes and shellfishes, Japan. Patent Appl. No. 55-24343, Feb., 27, 1980, Laid-open No. 56-121441, Sept. 24, 1981.

63. Grisolia, S., O'Conner, J. E., and Costell, M., L-carnitine and ammonia toxicity, Bull. Mem. Acad. R. Med. Belg. 139, 289, 1984.

64. Costell, M., Miguez, M. P., O'Conner, J. E., and Grisolia, S., Effect of hyperammonemia on the levels of carnitine in mice, <u>Neurology</u>, 37, 804, 1987.

65. Burtle, G. J. and Newton, G. L., Effect of carnitine on ammonia stressed channel catfish fingerling, Georgia

Aquaculture Report, Univ. of Georgia Coll. of Ag. and School of Forestry Cooperating, Athens, in press, 1991.

66. Burtle, G. J. and Newton, G. L., Effects of dietary carnitine supplements on growth and tissue lipid of fingerling channel catfish, submitted, 1991.

67. Carter, M., Carter, A. L., and Burtle, G. J., Butyrobetaine hydroxylase activity in channel catfish tissues, unpublished, 1991.

68. Carter, A. L., Stratman, F. W., Hutson, S. M., and Lardy, H. A., The role of carnitine and its esters in sperm metabolism, in <u>Proc. Virginia Lazenby O-Hara Biochem. Symp., Carnitine Biosynthesis, Metabolism, and Functions</u>, Frankel, R. A. and McGarry, J. D., Eds., Univ. Adelaide, Glen Osmond, Australia, 1979. Academic, New York, NY, 1980, 251.

69. Arduini, D., Rizzo, G., Mancinelli, R., Metro, D., Dell'Acqua, S., and Romanini, C., Effects of free fatty acids and L-carnitine on the spontaneous motility of rat isolated uterine horn, <u>Gynecol. Obstet. Invest.</u>, 24, 198, 1986.

70. Pesti, G. M., The nutrition of labile methyl group donors in broiler chickens, <u>Proc. 1989 Georgia Nutr Conf. for the Feed Industry</u>, Univ. of Georgia Ag. Exp. Sta., p. 145, 1989.

DISCUSSION OF THE PAPER

P. BORUM *(Gainesville, FL)*: Was the decrease in lipid in swine a decrease in muscle lipid or a decrease in adipose deposits?

NEWTON: You are referring to the work at Kansas State. Dr. Jim Nelssen, who was the principal investigator for the work is in the audience and may want to comment. The determinations were made on the whole animal. Pigs were sacrificed, the whole body was ground, mixed, sampled and analyzed such that any compartmentalization of the lipid reduction was not addressed. I believe Dr. Nelssen made a comment concerning procedures, but I do not remember specifically what it was.

L. BIEBER *(East Lansing, MI)*: Have you considered the relationship of the growth promoting effects of L-carnitine and the studies that show BGH increases the lean mass to fat ratio?

NEWTON: It is my understanding that supplemental growth hormone has very little influence on the body composition of 3 to 8-week-old pigs, such as those used in the carnitine research. Dr. Nelssen has been involved in studies with both compounds and may want to comment.

J. NELSSEN *(Manhattan, KS)*: The fat lowering response to supplemental carnitine is much greater than observed for PST (porcine somatotropin). PST has little effect on body composition of nursery pigs.

V. BOBYLAVA *(Guarri, Italy)*: What effect does the route of administration have on the response to supplemental carnitine? What are the differences between i.v. and oral treatment?

NEWTON: There are reports of both i.v. (injection or infusion) and oral administration of carnitine in dairy cattle, pigs, rats, mice, guinea pigs, horses, humans and probably rabbits and goats. It would appear that there are differences depending upon the route of supplementation, O'Conner and others [1] found differences in response to i.p., i.v., i.m. and s.c. carnitine. I am unaware of any report of an experiment where there was a direct comparison between oral and injectable carnitine. Speculation suggests that an oral dose would have to be larger than an infused dose to achieve the same response, due to possible digestive tract losses which could vary with species. Oral carnitine should enter the system more slowly than an injection, especially if it were mixed with the feed. Oral administration should also, depending upon the level fed, produce larger increases in digesta carnitine concentration and in portal carnitine concentration, than peripheral injection. Your question does not have a definitive answer at this time, however it could be addressed, if a suitable model and response criteria were selected, by comparing the response to peripherally infused, portally infused and orally administered carnitine at levels to provide equivalent plasma concentrations.

1.   O'Conner, J.E., Costell, M., Miguez, M.P., and Grisolia, S. Influence of the route of administration on the protective effect of L-carnitine against acute hyperammonemia, <u>Biochem. Pharmacol.</u> 35, 3173, 1986.

CARNITINE-MEDIATED ATTENUATION OF ETHANOL METABOLISM

Dileep S. Sachan, D.V.M., Ph.D.
Department of Nutrition, College of Human Ecology and
Agricultural Experiment Station, University of Tennessee,
Knoxville, TN 37996-1900, U.S.A.

The role of carnitine in fatty acid metabolism is well established and literature continues to grow on carnitine associated pathophysiologic conditions. Several studies have reported on the effects of carnitine on ethanol-induced alterations of lipid metabolism (1-4), however, studies pertaining to the effects of carnitine on metabolism of ethanol itself are just beginning to emerge (4-8). The objective of this report is to summarize our studies on the effects of carnitine, as a dietary supplement, on ethanol metabolism.

I. EFFECTS OF SUPPLEMENTARY CARNITINE ON BLOOD-ETHANOL CONCENTRATIONS

During the course of a previous study it was observed that adult male rats fed a liquid ethanol diet supplemented with D,L-carnitine were relatively inactive compared to those fed the diet without carnitine supplement. An examination of this observation revealed that D,L- carnitine supplemented (CS) rats maintained higher blood-ethanol concentrations (BEC) than the nonsupplemented (NS) rats (Table 1 and ref. 5). A later study showed the same effect with L-carnitine (Table 1). The BEC were related to blood nonesterified carnitine concentrations which were in turn related to the various levels of D,L-carnitine added to the regular Purina rat chow (Table 2 and ref. 5 for details).

TABLE 1

Effects of Carnitine (CNE) Supplemented Diets on Blood-Ethanol Concentrations (BEC) in Rats

Treatment diet CNE			BEC (mg/dl)	
			NS	CS
Chronic	LED	1.0% D,L	118.6 ± 9.8(5)[a]	237.4 ±19.9(5)[b]
Acute	PRC	1.0%, D,L	84.0 ± 4.5(4)[a]	126.0 ± 6.5(4)[b]
Acute	PRC	0.5%, L	82.5 ± 5.2(5)[a]	126.0 ± 1.6(5)[b]

LED = Liquid ethanol diet fed for 8 weeks.
PRC = Purina rat chow fed for 6/7 days, BEC 2 hr post- ethanol.
Values are mean ± SEM and those bearing same superscript are not
statistically different (p>0.05). Numbers in parenthesis represent no.
of rats/ group. (See Reference 5 for details.)

TABLE 2

Carnitine and Ethanol Concentrations in Blood of Rats Fed Diets Supplemented With Various Levels of D,L-Carnitine

Levels of carnitine	Blood concentrations	
	Carnitine (nm/ml)	Ethanol (mg/dl)
None	28.4 ± 4.3[a]	77.0 ± 11.0[a]
0.25%,w/w	81.3 ± 7.0[b]	128.0 ± 5.0[b]
0.50%,w/w	125.2 ± 9.0[c]	137.5 ± 4.5[b]
0.75%,w/w	130.2 ± 3.6[c]	177.0 ± 4.0[c]
1.00%,w/w	160.5 ± 5.6[d]	213.5 ± 10.5[c]

Values are mean ± SEM for 5 rats fed Purina rat chow for
5 days before testing. Values within a column bearing same
superscript are not significantly different (p > 0.05).
(See Reference 5 for details.)

Time course of BEC following a single oral dose of ethanol showed that BEC of the male rats fed CS Purina rat chow were significantly higher than those of the NS rats from hours 1 through 8 of post-ethanol administration (PEA). Since BEC in the CS and NS rats were not significantly different during the first 60 minutes of PEA, it was concluded that carnitine-mediated attenuation of ethanol clearance was most likely due to changes in ethanol metabolism (Figure 1 and Reference 6).

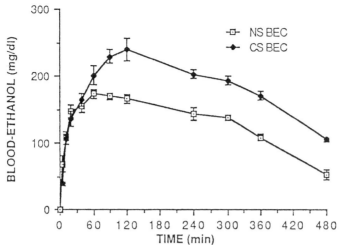

Figure 1. Time course of blood-ethanol concentrations (BEC) in rats fed Purina Chow, without (NS) and with D,L-carnitine supplementation (CS).

II.   EFFECTS OF L-CARNITINE ON PORTAL AND SYSTEMIC BEC:

While the above studies clearly demonstrated attenuation of ethanol clearance by D,L-carnitine (6) it was not clear if the effect was due to the D-isomer of carnitine present in the racemic mixture. Furthermore, it could not be said with certainty that altered absorption of ethanol from the gastrointestinal tract (GIT) was not at all a factor in raising the BEC of the CS rats. These concerns were addressed by feeding various supplementary levels of L-carnitine and determining portal and systemic BEC following a single oral dose of ethanol (7). Mature male rats were fed Purina rat chow supplemented with four levels of L-carnitine for 10 days. The concentrations of carnitine in the blood had attained steady state. Following general anaesthesia (Ketaset), each animal was fitted with a catheter in the portal vein and in the posterior venecava for serial blood sampling. Also, duodenal canula was placed through the stomach wall for ethanol infusion. A dose (3g/ Kg bw) of ethanol as 13% solution was intraduodenally infused (@ 1 ml/min) and 20 μl blood samples were collected simultaneously from the portal vein and posterior venecava at 5, 10, 15, 20 and 30 minute intervals. The BEC in the portal blood of the CS rats were not significantly different from those of the NS rats at all time points. However, the systemic BEC of the CS rats were significantly higher than those of the NS rats at corresponding time points (see Table 3 for one set of data and ref. 7 for additional details). From this data it was concluded that L-carnitine caused a rise in BEC, and that the effect was not due to enhanced ethanol absorption from the small intestine, but to attenuation of ethanol metabolism.

TABLE 3

Portal and Systemic Blood-Ethanol Concentrations Following
Intraduodenal Ethanol Infusion in Rats Fed NS and CS Diets

Post-ethanol infusion	Blood-ethanol concentrations (mg/dl)			
	Portal		Systemic	
(Min)	NS (4)	CS (4)	NS (4)	CS (4)
5	591 ± 33[a]	613 ± 42[a]	115 ± 25[b]	256 ± 44[c]
10	875 ± 74[a]	953 ± 91[a]	187 ± 12[b]	308 ± 40[c]
15	578 ± 77[a]	664 ± 56[a]	218 ± 12[b]	332 ± 25[c]
20	495 ± 99[a]	487 ± 56[a]	239 ± 27[b]	337 ± 26[c]
30	342 ± 51[a]	532 ± 52[a]	168 ± 29[b]	263 ± 3[c]

NS = nonsupplemented, CS = 0.25% L-carnitine supplemented Purina rat
chow fed for 10 days. Values are mean ± SEM and those bearing common
superscript are not significantly different ($p > 0.05$). (See Reference
7 for additional values and details.)

III. EFFECTS OF CARNITINE ON BEC FOLLOWING INTRAPERITONEAL ETHANOL ADMINISTRATION:

These experiments were designed to further assess the
effects of L-carnitine on ethanol metabolism when ethanol
was parenterally administered. A group of mature male rats
fed Purina rat chow were given an i.p. dose (0.2 g/Kg bw) of
ethanol (50% solution). Serial blood samples (tail vein)
were then collected for ethanol determination. Three days
later the same rats were placed on 0.5% L-carnitine
supplemented Purina chow. After 3 days of feeding CS chow,
they were administered an i.p. dose of ethanol and serial
blood samples were collected and analyzed for BEC. The BEC
in the CS period were significantly higher than in the NS
period (Table 4). On the other hand, rates of blood-ethanol
clearance during the CS period were significantly lower (5.5
mg/dl/hr) than during the NS (9.1 mg/dl/hr) period. These
results confirmed carnitine-mediated attenuation of ethanol
metabolism and removed any residual concern about the effect
being due to altered gastric emptying.

TABLE 4

Blood-Ethanol Concentrations Following I.P. Ethanol Dose in Rats Fed 0.5% Supplementary L-Carnitine for 3 Days

Post-ethanol administration (minutes)	Blood-ethanol concentrations (mg/dl)	
	NS	CS
5	203.2 ± 5.6^a	211.0 ± 4.2^a
10	194.8 ± 3.5^a	207.0 ± 7.3^a
20	179.8 ± 8.1^a	208.0 ± 5.0^b
40	163.0 ± 6.3^a	208.8 ± 5.0^b
80	155.8 ± 7.2^a	199.8 ± 4.4^b
160	141.2 ± 5.8^a	196.2 ± 6.1^b
320	134.4 ± 6.6^a	182.8 ± 7.5^b
480	118.2 ± 1.6^a	174.2 ± 7.5^b

There were 5 rats in each group. All animals were fed with (CS) or without (NS) 0.5% supplementary L-carnitine for 3 days prior to intra-peritoneal ethanol injection. The values are group means $\pm$ SEM and those in a line bearing the same superscript letter are not statistically different ($p \geq 0.05$).

IV. EFFECTS OF L-CARNITINE ON ETHANOL-INDUCED SLEEPING TIME:

The physiological consequences of BEC were assessed by monitoring the ethanol-induced sleeping time (loss of righting reflex) in male rats given, i.p., ethanol (2.5 g/Kg b.w.) following 0.5% L-carnitine supplementation. The sleeping time was significantly longer in CS than in NS rats, however, the induction of sleeping time was not significantly different (Table 5). Similar increases in ethanol-induced sleeping time have been recorded in the male mice fed CS (40.9 sd 11.9) compared to those fed NS (25.1 sd 7.9 min) Purina mouse chow.

TABLE 5

Effect of L-Carnitine on Ethanol-Induced Sleeping Time

Parameter	NS	CS
Induction time (min)	2.45 ± 0.4[a]	2.64 ± 0.6[a]
Sleeping time (min)	17.85 ± 4.9[a]	45.42 ± 2.9[b]

NS = nonsupplemented, CS = 0.5% L-carnitine supplemented
Values are mean ± SD for 4 male rats/ group and those with
same superscript letter are not different (p > 0.05).

V.   CARNITINE ATTENUATED ETHANOL METABOLISM AT VARIOUS DOSES OF ETHANOL:

The objective of this series of experiments was to determine if carnitine-mediated attenuation of ethanol metabolism was consistent for low as well as high doses of ethanol. Ten male rats weighing 450-500 g were fed NS or 0.5% L- CS Purina rat chow for 90 days. Starting with the lowest dose, five different, i.p., doses of ethanol were administered daily beginning on day 91 and ending on day 95. Each day following ethanol dosing, blood samples were collected from the tail vein for ethanol determination. The rates of blood-ethanol clearances were slower in the CS than in the corresponding NS rats at the doses of 0.05, 0.1, 0.2, 0.4 or 0.8 g/Kg bw, respectively. However, at the highest dose (0.8 g ethanol/ Kg bw) the 0.5% carnitine was not as efficacious as it was at the lower doses of ethanol (Table 6). It was concluded that carnitine mediated attenuation of ethanol metabolism was effective at low as well high doses of ethanol.

TABLE 6

Effects of Carnitine on Pharmacokinetics of Ethanol
at Various Doses of Intraperitoneal Ethanol Administration

ETOH dose	NS		CS	
(g/Kg bw)	C_o	K_{el}	C_o	K_{el}
0.05	192.2	24.0	193.7	9.0
0.10	188.9	14.4	152.8	5.4
0.20	215.0	15.6	245.7	9.0
0.40	237.6	9.6	266.5	5.4
0.80	250.4	16.2	301.3	13.2

NS = Nonsupplemented, CS = 0.5% L-carnitine supplemented
C_o = Zero time BEC (mg/dl)
K_{el} = Blood ethanol elimination (mg/dl/hr)

VI. LAG TIME REQUIRED FOR CARNITINE-MEDIATED ATTENUATION OF ETHANOL METABOLISM IN RATS:

It had been observed that simultaneous oral
administration of carnitine with ethanol did not affect BEC.
Therefore, the objective of these experiments was to
determine how long in advance carnitine must be administered
to bring about attenuation of ethanol metabolism seen in
earlier studies.

A. ORAL ADMINISTRATION:

Eight male rats (400-600 g bw) fed Purina chow were
randomly divided into two groups. One group was gavaged
with water (NSG) and the other group was gavaged with L-
carnitine (CSG) solution (200 mg/ml) @ 400 mg L-carnitine/
Kg bw which was equivalent to the daily dose of carnitine
received from the diet. A single oral dose (3g/Kg bw) of
ethanol was administered 24 hrs and 56 hrs after the
carnitine gavage. Serial blood samples were collected after
each ethanol dose and analyzed for BEC. Four hours after
the second ethanol dose the animals were killed and their
livers were collected. There were no significant
differences in the BEC of NSG and CSG until 8 hrs after the
first ethanol dose i.e. 32 hrs after the first carnitine
gavage. The higher BEC in the CSG rats were maintained
after a second dose of ethanol. The pharmacokinetic
parameters of blood ethanol during the 24-32 hr period after
the first carnitine dose are shown in Table 7. Sixty hrs
after the carnitine gavage and 4 hrs after the second dose

of ethanol, livers of the CS rats contained only 52% of the ethanol contained in the livers of NSG rats (see Table 8).

B. INTRAPERITONEAL ADMINISTRATION:

Ten male rats (400-600 g bw) fed Purina rat chow were divided into two groups, NSG and CSG. On day 1, both groups were given a dose of ethanol (0.2 g/ Kg bw, as 50% solution). The ethanol dose was immediately followed by a dose of L-carnitine (400 mg/ Kg bw) in solution (200 mg/ml) in CSG and with an equal volume of sterile normal saline in the NSG rats. On day 2, carnitine and normal saline (no ethanol) administration was repeated. On day 3, carnitine and placebo were given again and immediately followed by a dose of ethanol. Serial tail vein blood samples were collected after each dose of ethanol and analyzed for BEC. There was no effect of carnitine on BEC during first 24 hrs of treatment. However, 4 hrs after the second ethanol dose or third carnitine dose (52 hr from first carnitine dose) BEC of the CSG was significant higher than that of the NSG animals. The pharmacokinetic parameters of blood ethanol during the 48-56 hr period of these experiments are shown in Table 7.

C. INTRAVENOUS ADMINISTRATION:

Eight male rats (400-600 g bw) fed Purina rat chow were fitted with jugular catheters. After recovery, 4 rats were given, i.v., a dose of L-carnitine solution (400 mg/ ml) @ 400 mg/ Kg bw (CSG) and the other 4 rats were given sterile normal saline. Immediately following carnitine administration, all animals were given, i.v., 50% solution of ethanol (@ 0.2 g ethanol/ Kg bw). These doses of carnitine and ethanol were administered on second and third day in sequence. On the third day, 6 hr after the third dose of ethanol, or 54 hrs after the first carnitine dose, the rats were killed and livers removed for analyses. The BEC were not different between the two groups during the first 24 hr, however it tended to be higher in the CSG during the next 24 hr, and significantly so after 52 hr from the first injection of carnitine. The pharmacokinetic parameters of blood ethanol during the 48-52 hr period are shown in Table 7. Fifty four hrs after the first carnitine injection and 6 hrs after the third dose of ethanol, livers of CSG rats contained only 51% of ethanol contained in the livers of NSG rats (see Table 8).

From this data it is concluded that carnitine retards ethanol metabolism and that advance (32-54 hrs) presence of supplementary carnitine is essential for it to cause attenuation of ethanol metabolism, regardless of the route of administration. The lag time requirement and liver-ethanol differential suggests that carnitine is acting through a second factor which may in turn modulate enzyme(s) and/ or membranes.

TABLE 7

Lag Time for L-Carnitine Effect by Different Routes of L-Carnitine and Ethanol Administration

Route of Adm.	hrs PFCD	NSG		CSG	
		C_o	K_{el}	C_o	K_{el}
Oral	24-32	291.9	16.5	281.4	9.5
I.P.	48-56	182.1	10.4	186.2	4.3
I.V.	48-54	323.8	14.9	345.0	10.1

PFCD = psot first carnitine dose
NSG = Normal saline group, CSG = Carnitine solution group
C_o = Zero time BEC (mg/dl)
K_{el} = Blood ethanol elimination (mg/dl/hr)

TABLE 8

Liver-Ethanol Contents After Oral (p.O.) or Intravenous (i.v.) Ethanol Plus L-Carnitine Treatment

Time lapsed from adm. of 1st CNE Dose	Last ETOH dose	Liver-Ethanol (mg/ g) NSG	CSG
60 h (0.4g/Kg)	4 h (3g/Kg),P.O.	57.7 ± 6.3a	30.0 ± 7.4b
54 h (0.4g/Kg)	6 h (0.2g/Kg),I.V.	47.8 ± 8.1a	24.4 ± 3.5b

CNE = L-carnitine, ETOH = Ethanol
Values are group mean ± SEM for 4 rats and those with common superscript letter are not significantly different at $p > 0.05$.

VII.   CARNITINE ALTERED ETHANOL OXIDATION BUT NOT ENZYMATIC ACTIVITIES:

Mature male rats were fed NS or CS diets and a dose of ^{14}C-ethanol was administered. They were placed in the metabolic chambers and expired CO_2 was collected for 12 hrs and then analyzed for radioactivity. The CS rats expired significantly lower amounts of $^{14}CO_2$ than the NS rats indicating a decrease in ethanol oxidation caused by 0.5% L-carnitine (Table 9). The in vitro activities of hepatic ADH, MEOS and Catalase, were not significantly different in the NS and CS rats (Table 10). It was concluded that carnitine modulated ethanol oxidation via mechanism other than alterations in the activities of ethanol metabolizing enzymes (see ref. 8 for details).

TABLE 9

Effect of L-Carnitine Supplemented Diet on [1-^{14}C]-Ethanol Oxidation in the Rat[1]

Time PEA[2] (hrs)		Groups			
	NS		CS		
DPMs x 10^{-6}	% of dose	DPMs x 10^{-6}	% of dose	%ΔNS-CS
1	0.54	3.1	0.48	2.6	.5
2	1.21	6.9	1.02	5.6	1.3
3	1.90	10.8	1.60	8.8	2.0
4	2.82	16.1	2.34*	12.8	3.3
5	3.60	20.5	3.12**	17.1	3.4
6	4.82	27.4	4.35**	23.8	3.6
7	6.12	34.8	5.46**	29.9	4.9
8	7.97	45.4	7.27**	39.8	5.6
10	11.31	64.4	10.40*	57.0	7.4
12	15.00	85.4	13.73**	75.2	10.2

[1]Oxidation of ethanol is represented as expired $^{14}CO_2$.
Values are group means ± SEM (n=6) and those bearing asterisk(s) are significant (*) $p \leq 0.05$, (**) $p \leq 0.01$. See ref. 8 for details.
[2]PEA = Post-ethanol administration.
NS = Nonsupplemented, CS = L-carnitine supplemented.

TABLE 10

Effect of L-Carnitine Supplemented Diet[1] on Ethanol
and Aldehyde Metabolizing Enzymes in Rat Liver

	Groups	
Enzyme	NS	CS
ADH[2]	8.26 ± 0.45[a]	9.12 ± 0.78[a]
Catalase[3]	0.93 ± 0.07[a]	0.89 ± 0.03[a]
MEOS[4]	14.86 ± 1.89[a]	11.51 ± 0.92[a]
ALDH[2]	32.31 ± 4.00[a]	27.94 ± 3.41[a]

[1] 0.5% L-carnitine (inner salt) supplemented Purina Chow fed for
7 days (CS). NS=Nonsupplemented. See ref. 8 for details.
[2]Activity is expressed as nmoles of NAD^+ reduced/min/mg protein.
[3]Activity is expressed as nmoles of H_2O_2 reduced/min/mg protein.
[4]Activity is expressed as nmoles Acetaldehyde/min/mg protein.
All values are group means ± SEM (n=5) and those bearing the
same superscript in a row are not significantly different (p >0.05).

SUMMARY

L-Carnitine supplemented in the diet raised BEC in mature
male rats which was not due to altered ethanol absorption
from the GIT as evident from the BEC data generated
following enteral and parenteral ethanol administrations.
Ethanol-induced sleeping time was prolonged by carnitine.
Carnitine-mediated attenuation of blood-ethanol clearance
was carnitine-dose dependent and present at low and high
doses of ethanol administration. A lag time was needed for
the effect of carnitine on BEC, regardless of the route of
carnitine administration. Liver-ethanol concentrations
were inversely related to the BEC. Ethanol oxidation in the
intact rat was reduced by carnitine, however, the activities
of ethanol metabolizing enzymes were not altered. It is
concluded that carnitine retards ethanol metabolism.

ACKNOWLEDGEMENT

This work was supported by the grant (TN 745) from the
Tennessee Agricultural Experiment Station. The data
reported here are part of the graduate studies of Dr. Robert
Berger and Dr. Randy Mynatt. L-carnitine used in many of
these experiments was supplied by the Sigma Tau
Pharmaceuticals, Inc.

REFERENCES

1.   Hosein, E.A. and Bexton, B. Protective action of carnitine on liver lipid metabolism after ethanol administration to rats. <u>Biochem. Pharmacol.</u> 24: 1859, 1975.

2.   Sachan, D.S., Rhew, R.H. and Ruark, R.A. Ameliorating effects of carnitine and its precursors on alcohol-fatty liver. <u>Am. J. Clin. Nutr.</u> 39: 738, 1984.

3.   Rhew, T.H. and Sachan, D.S. Dose-dependent lipotropic effect of carnitine in chronic alcoholic rats. <u>J. Nutr.</u> 116: 2263, 1986.

4.   Adamo, S., Siliprandi, N., Di Lisa, F., Carrara, M., Azzurro, M., Sartori, G., Vita, G. and Ghidini, O. Effect of L-carnitine on ethanol and acetate plasma levels after oral administration of ethanol in humans. <u>Alcoholism</u> 12: 653, 1988.

5.   Berger, R. and Sachan, D.S. Elevation of blood-ethanol concentrations in carnitine supplemented rats. <u>Nutr</u>. <u>Rep. Inter.</u> 34:153, 1986.

6.   Sachan, D.S. and Berger, R. Attenuation of ethanol clearance by supplementary carnitine in rats. <u>Alcohol</u> 4:31, 1987.

7.   Berger, R. and Sachan, D.S. Effects of supplementary L-carnitine on blood and urinary carnitines and on the portal-systemic blood-ethanol concentrations in the rat. <u>J. Nutr. Biochem.</u> 2(7) 382, 1991.

8.   Mynatt, R.L. and Sachan, D.S. L-Carnitine retards ethanol oxidation without alterations in enzymatic activities. <u>Alcoholism: Clin. Exptl. Res</u>. In Review, 1991.

DISCUSSION OF THE PAPER

M. NOVAK *(Miami, FL)*: Are there any differences in the liver histology of control and experimental animals?

SACHAN: Histological examination of the livers was not carried out in the acute or single dose experiments, however, in the chronic ethanol study liver histology was characterized as being similar to that of the control.

T. GREWAY *(Nutley, NJ)*: If initial blood alcohol level is same in CS and NS groups, and liver activities of alcohol detoxicating enzymes are the same, what is the possible mechanism of carnitine increasing the blood concentrations of ethanol?

SACHAN: The mechanism is not known at this point. The liver of the CS rats does not retain as much ethanol as does the liver of the NS rats (Table 8). The speculation is that hepatic blood volume of CS rats may be altered and/or diffusion of ethanol across hepatocyte membrane be altered by carnitine.

C. HOPPEL *(Cleveland, Hts., OH)*: Does carnitine affect the disposition kinetics of i.v. ethanol and alter the dose dependent handling of ethanol?

SACHAN: Following an i.v. ethanol dose, ethanol elimination rate was lower in CS rats (Table 7) and liver ethanol levels were inversely related to the blood ethanol concentration (Table 8). Disposition kinetics at various doses of ethanol was not determined.

DESIGNING ACTIVE-SITE PROBES OF CARNITINE ACYLTRANSFERASES: POTENTIALLY POTENT REGULATORS OF LIPID METABOLISM

RICHARD D. GANDOUR
DEPARTMENT OF CHEMISTRY
LOUISIANA STATE UNIVERSITY
BATON ROUGE, LA 70803-1804

We are designing, preparing, and assaying conformationally rigid inhibitors of carnitine acetyltransferase (CAT), carnitine octanoyltransferase (COT), and carnitine palmitoyltransferase (CPT), which are enzymes that catalyze acyl transfer between carnitine and coenzyme A (CoA). The primary aim is to identify the topographical arrangement of the key recognition sites on these enzymes by comparing inhibition constants (K_i's) of a series of rigid, cyclic analogues. Our long-range goal is to elucidate the structure of the molecular interactions between carnitine and the active site in these enzymes. Identifying these molecular interactions will lead to a better understanding of the physiological chemistry and, more specifically, of the regulation of the enzymes.

Conformationally rigid inhibitors can probe the structure of enzymic binding sites. This idea originates in medicinal chemistry, e.g., hemicholiniums,[1] which are analogues of choline. We are using the concept of a rigid inhibitor's organizing a flexible enzyme (induced-fit model[2]) to design molecular probes of the topographies of the catalytic centers in carnitine acyltransferases. We are preparing conformationally rigid analogues of reaction intermediates that we have proposed for carnitine-acyltransferase-catalyzed reactions. These analogues have groups anchored to a rigid molecular framework in a well-defined stereochemistry. The magnitude of K_i depends on how well the structure of the inhibitor complements the topography of the catalytic center. Because many enzymes use CoA as a substrate, we initially focus on designing carnitine analogues to ensure that our inhibitors will be specific for carnitine-dependent enzymes.

I. POSSIBLE BINDING SITES IN CARNITINE ACYLTRANSFERASES.

In the absence of three-dimensional molecular structures of these enzymes, we approach the design problem by imagining how the enzymes will recognize the two reactants. Based on previous studies, the enzymes, especially CAT,[3] recognize trimethylammonium, carboxylate, and acyl on acylcarnitine.(Figure 1) CoA has many possible recognition points, one of which, based on early work of Chase,[4] is the 3'-phospho. Previously, we have proposed[5] that binding of both trimethylammonium and carboxylate are needed for chiral discrimination. This is incorrect. The enzyme needs to recognize only one of these ionic groups, because the location CoA and the stereocenter on carnitine are the other two points that are required for chiral discrimination between (*R*)- and (*S*)-carnitine or (*R*)- and (*S*)-acylcarnitine. [(*R*)-and (*S*)-carnitine ≡ L- and D-carnitine, respectively]

Possible Recognition Points for Acylcarnitine and CoA

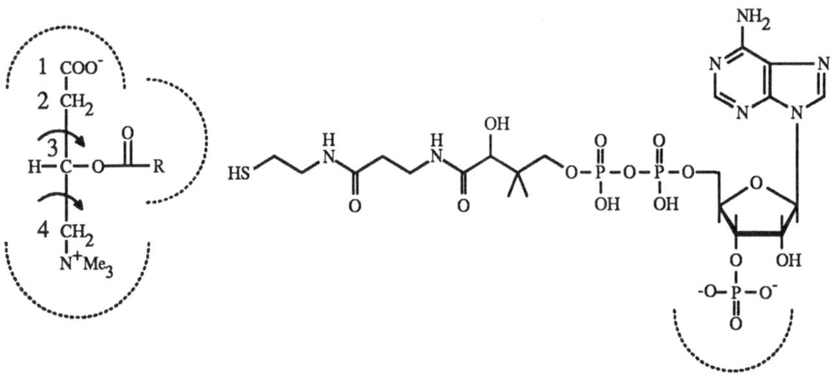

Figure 1. Possible recognition sites on carnitine acyltransferases for acylcarnitine and CoA. Hashed lines represent groups that when modified affect the binding.

II. CONFORMATIONAL ANALYSIS OF CARNITINE AND ACETYLCARNITINE

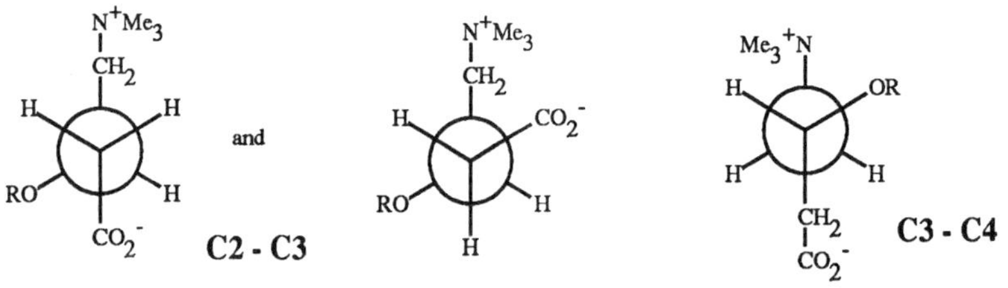

Method	**anti**	**gauche (-)**	**gauche (-)**	**R**
NMR	0.53	0.42	0.91	H
MM2	0.47	0.41	0.99	H
NMR	0.36	0.55	0.98	Ac
MM2	0.45	0.53	0.98	Ac

Figure 2. Fractional populations of conformations.

To design a rigid analogue, we must know which of nine possible conformations of carnitine is (are) bound to the enzyme. This question remains unanswered, but we have performed conformational analyses of carnitine and acetylcarnitine in other phases by single crystal X-ray diffraction,[5] NMR, and molecular mechanics (MM2).[6] Figure 2 shows the fractional population of conformations as determined by NMR and MM2. The other conformation(s) (not shown) comprise the total population (1.0). In solution, both compounds adopt a highly preferred gauche(-) conformation about the N-C4-C3-O torsion angle. In contrast, both exhibit a nearly equal preference for gauche(-) and anti about the C1-C2-C3-C4 torsion angle, with carnitine favoring anti and acetylcarnitine favoring gauche(-).

X-ray studies[5] of the zwitterions of carnitine (Figure 3) and acetylcarnitine (Figure 4) reveal similar results to the solution studies. Carnitine has a different conformation than acetylcarnitine. The conformation about C3-C4 is similar in both, but the conformation about C2-C3 changes from anti to gauche(-). Murray, Reed, and Roche[7] have labeled the conformation of carnitine as 'extended' and that of acetylcarnitine as 'folded'.

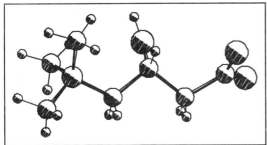

Figure 3. (above) Crystal Structure of Carnitine Zwitterion.

Figure 4. (right) Crystal Structure of Acetylcarnitine Zwitterion.

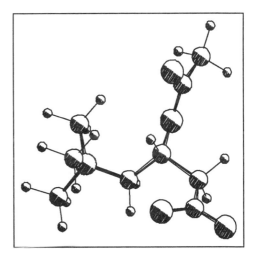

In summary, our conformational analyses of carnitine and acetylcarnitine show that carnitine prefers 'extended' or anti-gauche(-) and acetylcarnitine 'folded' or gauche(-)-gauche(-). The important points for inhibitor design are that: (a) the C3-C4 bond strongly prefers gauche(-), and (b) the C2-C3 torsion is equally populated in either the anti or gauche(-) conformation. We can lock the C3-C4 torsion in the gauche(-) conformation by formation of a ring and probably not lose much in binding to the enzyme. We are less certain as to whether or not to lock the conformation of C2-C3, although we favor the gauche(-) for mechanistic reasons.(see below)

III. TRANSITION-STATE AND REACTION-INTERMEDIATE ANALOGUES

Wolfenden[8] has pioneered the development of transition-state analogue inhibitors of enzymes. The idea is that enzymes bind transition structures or reaction intermediates more tightly than reactants or products. Molecules that resemble the structures of transition states or reaction intermediates, but are unreactive, will bind strongly to the enzyme. Many such analogue inhibitors of peptidases are quite potent. The general structural feature of these analogues is a tetrahedral carbon or phosphorous in place of the trigonal carbonyl carbon of a peptide bond. This replacement mimicks the tetrahedral intermediate in peptide hydrolysis, which is formally an acyl transfer from an amine to water. Carnitine acyltransferases should bind similar tetrahedral-intermediate analogues. Our goal is to design conformationally rigid, reaction-intermediate analogue inhibitors of CAT, COT and CPT. To do this requires a bioorganic mechanism for the reaction.

In the absence of bioorganic mechanistic studies on these enzymes, we have proposed[5] a mechanism (Figure 5), based on chemical model studies of O-to-S acyl transfer,[9] for acetyl transfer in CAT. We presume that a similar mechanism operates in COT and CPT. The principle of microscopic reversibility requires that the forward and reverse reactions have the same rate-limiting step. The tetrahedral intermediate contains both carnitine and CoA. In the forward direction, deprotonation of hydroxyl needs a base (Enz-B), while in the reverse direction, protonation of alkoxide needs an acid (Enz-BH). At the optimal pH (7.8) for reaction, about 10% of CoA is ionized. The step involving CoA does not require acid-base catalysis. Without detailed kinetic studies of the bioorganic mechanism, the identification of the rate-limiting step will remain a mystery.

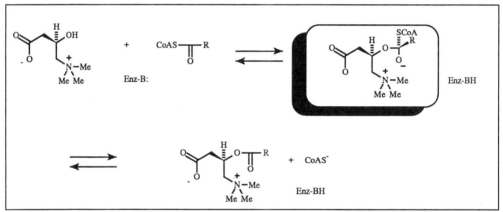

Figure 5. Proposed mechanism for acyl transfer between carnitine and CoA. The shadow box encases the putative tetrahedral intermediate. Enz-B(H) is a base and acid, probably a histidine.

The alkyl group (R) probably does not detach from the alkyl recognition site, especially in COT and CPT, during the acyl transfer. This fixed recognition site creates an additional chirality, viz., the tetrahedral intermediate. The enzymes also recognize the chirality at C3 of (acyl)carnitine during the transfer. A reaction-intermediate analogue must mimick both of these chiralities. These enzymes might prefer the *R* configuration at this center because the thiol group should approach the acyloxy from the less-congested side (carboxyl on carnitine 'folded' back). The direction of the attack is on the *Re* face of the ester, assuming that the acyloxyl group is in the most stable conformation.[5] (In

Figure 4, attack would occur from the middle of the right edge of the figure.) An additional benefit of this conformation is the electrostatic catalysis[10] that would result from the developing negative charge on the carbonyl oxygen that is near the positively charged trimethylammonium group.

IV. INHIBITOR DESIGN

We reason that a covalent bond can replace this electrostatic interaction.(Figure 6)

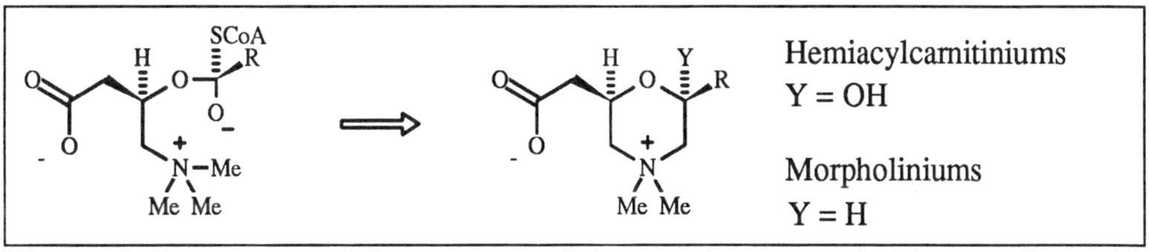

Figure 6. Comparison of the structure of the putative tetrahedral intermediate with the analogue inhibitor.

Knowing that the conformation about C3-C4 in carnitine is predominantly gauche(-), we have locked this conformation in a ring. Ease of synthesis and fewer conformations have influenced the choice of a six-membered ring, even though a seven- or eight-membered ring might be a better structural match to the putative tetrahedral intermediate. Conformational considerations in six-membered rings dictate that the carboxymethyl and the alkyl chain are cis diequatorial in the most stable conformation. We approach the syntheses of morpholinium inhibitors from two directions -- morpholiniums and hemiacylcarnitiniums (2-hydroxymorpholiniums). The latter inhibitors closely resemble the hemicholiniums, which inhibit acetylcholine synthesis by blocking the uptake of choline.[11] The hemiacylcarnitiniums have the same molecular formula as the corresponding acylcarnitines.

V. MORPHOLINIUMS

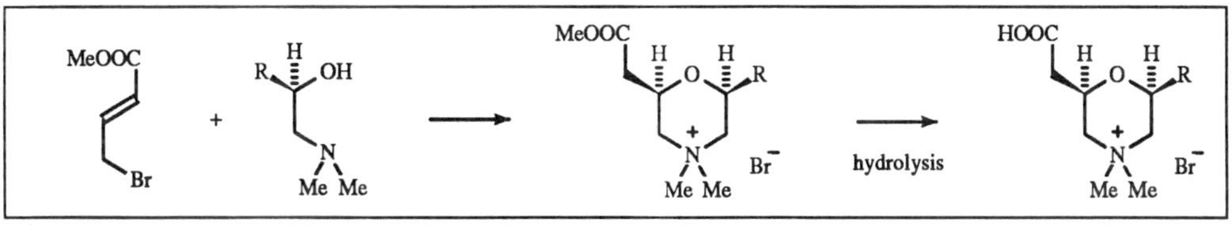

Figure 7. Scheme for the synthesis of morpholiniums.

The reaction of dimethylaminoalcohols with methyl 4-bromo-2-butenoate, followed by hydrolysis, produces the morpholiniums.(Figure 7) Alkylation of the amine occurs first followed stereoselective cyclization to the morpholinium ring occurs in a separate step.

We have prepared four compounds, **1** - **4**, in this class.[12](Figure 8) Reactions of 2-(*N,N*-dimethylamino)ethanol (R = H) and 1-(*N,N*-dimethylamino)-2-propanol (R = CH$_3$) with the bromoalkenoate give **1** and **2**, respectively. Compound **2** forms exclusively as the racemate of the cis diastereomer. We have made *meso*-2,6-bis(carboxymethyl)-4,4-dimethylmorpholinium, **4**, from condensation of sodium (*R*)-norcarnitine[13] (R = CH$_2$COONa) and the bromoalkenoate followed by hydrolysis of the half ester, **3**. Single crystal X-ray analysis has verified the structure of the diequatorial meso diacid. The solid-state structure of **4** displays meso symmetry, with a crystallographic mirror plane containing the O and N atoms of the ring. Hence, we call it a 'Siameso' inhibitor because of its morphological similarity to Siamese twins.

These compounds, conformationally rigid analogues of the tetrahedral intermediate proposed for acetyl transfer in CAT, moderately inhibit the pigeon-breast enzyme.(Figure 8) Pigeon-breast CAT binds both enantiomers of carnitine (*R*, K_m = 120 μM; *S*, K_i = 173 μM) and acetylcarnitine (*R*, K_m = 350 μM; *S*, K_i = 256 μM) equally well, but only the (*R*)-enantiomers are substrates.[14,15] (CATs from other sources show chiral discrimination of substrates.[16]) These analogues must have the same relative configuration as (*R*)-carnitine, if CAT stereoselectively binds them as it binds the tetrahedral intermediate. The data for competitive inhibition in Figure 8 support the idea that the enzyme recognizes the chirality of the inhibitors. Of this series **4** binds most strongly, with a K_i half that of the racemic compounds, **1** and **2**. Every molecule of **4** has one side with the correct configuration of (*R*)-carnitine. This two-fold improvement in binding for **4** suggests that CAT is selectively binding one configuration of these inhibitors. Compound **3** does not bind well because of the increased size or the decreased polarity of the ester or both.

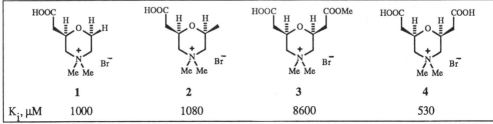

Figure 8. Morpholinium inhibitors of pigeon breast CAT.

The key features of these inhibitors are rigidity and similarity to the tetrahedral intermediate. Rigidity in the inhibitor reduces binding because only the enzyme can adjust, but rigidity is essential for identifying the topographical arrangement of recognition points on the enzyme, and the conformation of the substrate fragment of the tetrahedral intermediate. For example, the N-C4-C3-O torsion angle in the inhibitors is locked in the gauche(-) conformation, which we predict as the conformation for (acyl)carnitine bound to CAT. The inhibition by these compounds and the hemiacylcarnitiniums (see below) supports this prediction.

VI. HEMIACYLCARNITINIUMS

For the synthesis of carnitine analogues of 2-hydroxymorpholiniums, we require grams of norcarnitine. We make twenty-grams batches of norcarnitine by demethylating carnitine in high yield.[13] We then make the methyl ester of norcarnitine, which reacts with bromomethyl ketones to produce the morpholinium esters, **5**, which when hydrolyzed yield hemiacylcarnitiniums.(Figure 9) The highly favored ring closure only gives an axial hydroxyl because of the anomeric effect, the gauche effect, and steric effects.

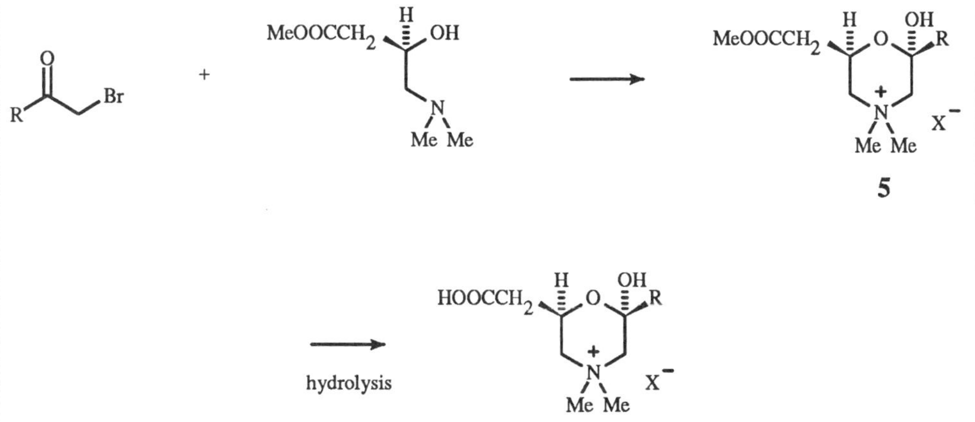

Figure 9. Scheme for the synthesis of hemiacylcarnitiniums.

The hemiketal carbon of hemiacylcarnitiniums and the hemiorthothioester carbon of the putative tetrahedral intermediate are chiral. The $2S$ configuration of a hemiacylcarnitinium is the same relative configuration as the proposed tetrahedral intermediate. We expect that of the four possible stereoisomers only the $2S,6R$ enantiomer will inhibit. Because the hydroxy is more stable in the axial position, formation of the hemiketal gives only one pair of enantiomers, $(2R,6S:2,6R)$. This stereoselective reaction excludes the formation of the $(2S,6S:2R,6R)$ pair, which have equatorial hydroxys. Unfortunately, we cannot use this class of inhibitors to determine the chirality of the reaction center by comparing the K_i's of the four stereoisomers. We are designing new morpholiniums to test the chirality of the tetrahedral intermediate.

We have prepared four hemiacylcarnitiniums (Figure 10) -- hemiacetyl-[17] (HAC), hemipropanoyl-[18] (HPrC), hemibenzoyl-[19] (HBC), and hemipalmitoyl-[20] (HPC), all of which inhibit carnitine acyltransferases.(Table 1) Chiral HAC binds to pigeon breast CAT twice as well as carnitine and six-fold better than acetylcarnitine. Chiral HAC binds 3.6-fold better than the racemate. This chiral selectivity demonstrates chiral recognition of the inhibitor, because this CAT binds both enantiomers of carnitine and acetylcarnitine equally well. Because HAC is a competitive inhibitor of both, we suggest that it occupies the same site as carnitine and acetylcarnitine. When HAC binds, the enzyme is in a conformation that recognizes the chirality, presumably the same conformation the enzymes adopts during catalysis. Chiral HAC strongly inhibits rat liver peroxisomal and rat heart mitochondrial CATs. Chiral HPrC inhibits pigeon breast CAT about 30% less than chiral HAC. Chiral HBC weakly inhibits pigeon breast CAT, but modestly inhibits the rat liver peroxisomal enzyme. HPC has no effect on pigeon breast CAT. These last examples resemble the pattern of chain length specificity of pigeon breast CAT for acylcarnitines as substrates.[15]

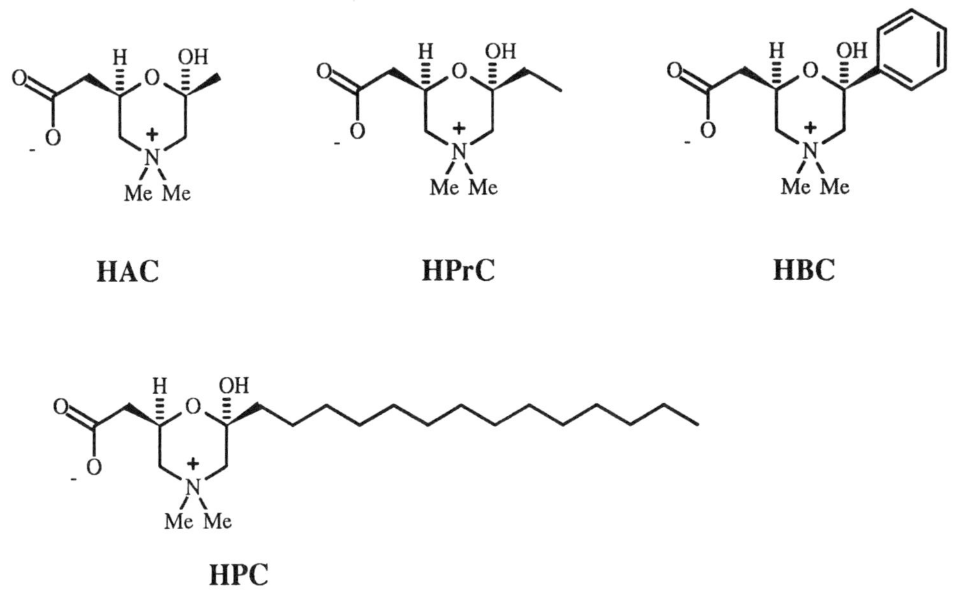

Figure 10. Structures of hemiacylcarnitiniums.

TABLE 1. Inhibition Constants (K_i, μM) of Hemiacylcarnitiniums.

Enzyme	HAC	HPrC	HBC	HPC
CAT, pigeon breast	59	198	3500	no effect
CAT, rat liver perox.	69	no assay	1112	no assay
CAT, rat heart mito.	114	no assay	2857	no assay
COT, rat liver	no assay	no assay	3700	no assay
CPT$_i$, rat liver	no effect	no assay	no effect	1.6

Racemic HPC strongly inhibits CPT$_i$, binding 9-fold better than (*R*)-palmitoylcarnitine (K_m = 14 μM). Chiral HPC should bind more tightly. HAC has no effect on CPT$_i$. The assay conditions contain saturating amounts of second substrate, so that the (apparent) K_i represents binding to the enzyme-CoA binary complex. In addition, purified CPT$_i$ contains Tween 20 at a concentration exceeding the critical micelle concentration. In spite of the nonideal conditions, the assays give reproducible Dixon plots that show competitive inhibition.[20]

VII. OTHER INHIBITORS OF PURIFIED CPT$_i$.

Strong inhibitors of purified CPT$_i$ (also called CPT-II or CPT-B) need only a long alkyl chain and a dimethylammonium ion.[20] (Figure 11) Racemic HPC competes with (R)-carnitine, palmitoyl-CoA, and (R)-palmitoylcarnitine with K_i's of 5.1, 21.5, and 1.6 μM, respectively. The differences in the K_i's represent binding to three different binary enzyme complexes -- CPT$_i$-PlmCoA, CPT$_i$-Cn, and CPT$_i$-CoA. HPC is uncompetitive with CoA for CPT$_i$-PlmCn. Palmitoylcholine and zwittergent 16 inhibit purified CPT$_i$ nearly as well as (R)-palmitoylcarnitine. The strong binding of palmitoylcholine compared with (R)-palmitoylcarnitine suggests that carboxylate contributes virtually nothing to the binding. All of these inhibitors bind more tightly to purified CPT$_i$ than 2-bromopalmitoyl-CoA (I_{50} = 353 μM) and tetradecylglycidyl-CoA (I_{50} = 60-80 μM). Racemic HPC is the most potent reversible synthetic inhibitor of purified CPT$_i$ to date.

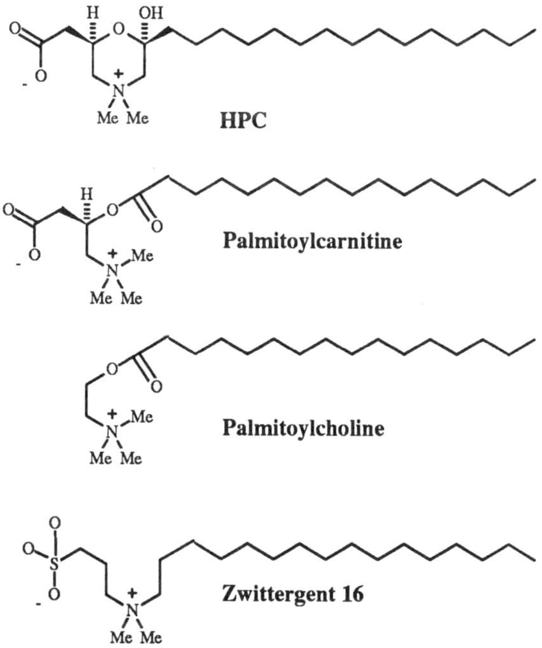

	Cn	PlmCoA	PlmCn
HPC	5.1	21	1.6
Palmitoylcarnitine	14	8	
Palmitoylcholine	16	10	
Zwittergent 16	46	64	483

Figure 11. Apparent K_i's (μM) of long-chain alkylammoniums for purified CPT$_i$. Cn = (R)-carnitine, PlmCoA = palmitoyl-CoA, PlmCn = (R)-palmitoylcarnitine.

We state the claim with care because of the many inhibitors of CPT's (see citations in ref. 20) and the number of different CPT's. (see chapter by S. V. Pande in this book) These efficacy of these inhibitors varies with certain preparations of CPT or with assay conditions -- purified protein, intact mitochondria, peroxisomes or microsomes. The location of malonyl-CoA, the physiological inhibitor of CPT activity, appears to be separate from the active site. (see chapters by J. D. McGarry and C. L. Hoppel in this book) We construct our inhibitors to target the active site.

VIII. SUMMARY AND CONCLUSIONS.

After conformational studies on carnitine and acetylcarnitine, and an analysis of a possible mechanism for acyl transfer, we have designed and synthesized effective inhibitors of CAT and CPT. These morpholinium derivatives provide a rigid framework from which to anchor the molecular fragments of the reaction. The activity of these inhibitors attests to the design rationale that the gauche(-) conformation about C3-C4 can be locked. Studies with these inhibitors directly address the question of the topographical arrangement of the key recognition sites for acylcarnitines in these enzymes. Because the inhibitors are rigid there is no question that their structural integrity is maintained on binding. They bind more strongly than the corresponding acylcarnitines, and according to the simplest interpretation of the data, they bind competitively at the active site. These results suggest that the inhibitors mimick well the spatial arrangement of the reacting molecules. HPC is currently the best inhibitor of purified CPT_i. We are working to verify that the stereochemistry of HAC and HPC are optimal by designing a new class of morpholiniums, in which we can prepare all four stereoisomers. We are also preparing an inhibitor for COT. The ultimate goal is to add a CoA fragment to the framework in order to determine the location and recognition pattern of the CoA binding site.

ACKNOWLEDGEMENTS

I gratefully thank my collaborators at LSU -- William J. Colucci, Noelle L. Blackwell, Stanhope P. Turnbull, Terry C. Stelly, Robert Hendricks, James W. Roche, Frank R. Fronczek -- for the structural and synthetic work. I especially thank my collaborators at other locations -- Linda J. Brady (U. of Minnesota), Paul S. Brady (U. of Minnesota), Eric P. Brass (Case Western U.), Loran L. Bieber (Michigan State U.), Chang Chung (Michigan State U.), and Rona P. Ramsay (VA - UCSF) -- for performing the enzymatic assays. I thank Sigma-Tau Pharmaceutical, Inc. for generous gifts of (*R*)-carnitine. I acknowledge the American Heart Association, Louisiana Chapter for initial support, Monsanto Corp. for an unrestricted grant, and the NIH through grant GM42016 for current financial support.

REFERENCES

1. Long, J. P., Schuler, F. W., A new series of cholinesterase inhibitors, *J. Pharm. Assoc. (Sci. Ed.),* 43, 79, 1954.

2. Koshland, D. E., Jr., Application of a theory of enzyme specificity to protein synthesis, *Proc. Natl. Acad. Sci. USA,* 44, 98, 1958.

3. Fritz, I. B., Shultz, S. K., Carnitine acetyltransferase. II. Inhibition by carnitine analogues and by sulfhydryl reagents, *J. Biol. Chem.,* 240, 2188, 1965.

4. Chase, J. F. A., pH-Dependence of carnitine acetyltransferase activity, *Biochem. J.,* 104, 503, 1967.

5. Gandour, R. D., Colucci, W. J., Fronczek, F. R., Crystal structures of carnitine and acetylcarnitine zwitterions. A structural hypothesis for mode of action, *Bioorg. Chem.* 13, 197, 1985.

6. Colucci, W. J., Gandour, R. D., Mooberry, E. S., Conformational analysis of charged flexible molecules in water by application of a new Karplus equation combined with MM2 computations: Conformations of carnitine and acetylcarnitine, *J. Am. Chem. Soc.,* 108, 7141, 1986.

7. Murray, W. J., Reed, K. W., and Roche, E. B., Conformation of carnitine and acetylcarnitine and the relationship to mitochondrial transport of fatty acids, *J. Theor. Biol.,* 82, 559, 1980.

8. Wolfenden, R., Transition-state affinity as a basis for the design of enzyme inhibitors, in *Transition States of Biochemical Processes*, Gandour, R. D. and Schowen, R. L., Eds., Plenum, New York, 1978, chap. 15.

9. Hupe, D. J., Jencks, W. P., Nonlinear structure-reactivity correlations. Acyl transfer between sulfur and oxygen nucleophiles, *J. Am. Chem. Soc.,* 99, 451, 1977.

10. Asknes, G., Prue, J. E., Kinetic salt effects and mechanism in the hydrolysis of positively charged esters, *J. Chem. Soc.,* 103, 1959.

11. Sterling, G. H., Doukas, P. H., Ricciardi, Jr., F. J., Biedrzycka, D. W., O'Neill, J. J., Inhibition of high-affinity choline uptake and acetylcholine synthesis by quinuclidinyl and hemicholinium derivatives, *J. Neuorchem.,* 46, 1170, 1986.

12. Colucci, W. J., Gandour, R. D., Fronczek, F. R., Brady, P. S., Brady, L. J., A 'Siameso' inhibitor: chiral recognition of a prochiral bilaterally symmetric molecule by carnitine acetyltransferase, *J. Am. Chem. Soc.* 109, 7915, 1987.

13. Colucci, W. J., Turnbull, Jr., S. P., Gandour, R. D., Preparation of crystalline sodium norcarnitine. An easily handled precursor for the preparation of carnitine analogues and radiolabeled carnitine, *Anal. Biochem.* 162, 459, 1987.

14. Chase, J. F. A., Tubbs, P. K., Some kinetic studies on the mechanism of action of carnitine acetyltransferase, *Biochem. J.*, 99, 32, 1966.

15. Chase, J. F. A., The substrate specificity of carnitine acetyltransferase, *Biochem. J.*, 104, 510, 1967.

16. Colucci, W. J., Gandour, R. D., Carnitine acetyltransferase: a review of its biology, enzymology, and bioorganic chemistry, *Bioorg. Chem.*, 16, 307, 1988.

17. Gandour, R. D., Colucci, W. J., Brady, P. S., Brady, L. J., Active-site probes of carnitine acetyltransferase. Inhibition of carnitine acetyltransferase by hemiacetylcarnitinium, a reaction intermediate analogue, *Biochem. Biophys. Res. Commun.*, 183, 735, 1986.

18. Blackwell, N. L., Fronczek, F. R., Gandour, R. D., (2*S*,6*R*)-6-ethyl-2-hydroxy-4,4-dimethylmorpholinium bromide, *Acta Crystallogr., Sec. C,* C45, 1089, 1989.

19. Blackwell, N. L., Fronczek, F. R., Gandour, R. D., (2*S*,6*R*)-6-carboxymethyl-2-hydroxy-4,4-dimethyl-2-phenylmorpholinium chloride hemihydrate, *Acta Crystallogr., Sect. C,* C46, 1164, 1990.

20. Gandour, R. D., Colucci, W. J., Stelly, T. C., Brady, P. S., Brady, L. J., Hemipalmitoylcarnitinium, a strong competitive inhibitor of purified hepatic carnitine palmitoyltransferase, *Arch. Biochem. Biophys.*, 267, 515, 1988.

DISCUSSION OF THE PAPER

F. Di LISA *(Italy)*: Does your carnitine derivative interfere with the activity of carnitine translocase?

R. GANDOUR: We don't know because we only had enough of the compound to test with the purified enzyme.

T. GREWAY *(Nutley, NJ)*: What were the K_m's for the substrates measured under reaction conditions that HPC gave a K_i of 1.6 μM? For palmitoyl-CoA, I get 15 μM.

R. GANDOUR: Our K_m's (μM) are: carnitine (100-200) and palmitoyl-CoA (30)[1].

J.D. MCGARY *(Dallas, TX)*: Does HPC gain access to CPT II in intact mitochondria?

R. GANDOUR: We would also like to know the answer to that question. Quite recently, we have overcome the synthetic difficulties in the preparation of HPC. We expect to have enough HPC for many assays.

1.   Gandour, R.D., Colucci, W.J., Stally, T.C., Brady, P.S., Brady, L.J., Hemipalmitoylcarnitinium, a strong competitive inhibitor of purified hepatic carnitine palmitoyltransferase, *Arch. Biochem. Biophys.*, 267, 515, 1988.

SITES OF ACTION OF CARNITINE: AN OVERVIEW

IRVING B. FRITZ
BANTING AND BEST DEPARTMENT OF MEDICAL RESEARCH
UNIVERSITY OF TORONTO, TORONTO, ONTARIO M5G 1L6, CANADA.

I. ACTIONS OF CARNITINE ON METABOLIC PROCESSES MEDIATED BY LONG-CHAIN CARNITINE ACYLTRANSFERASES

In a review on carnitine in 1957 by Fraenkel and Friedman[1], the authors reported that "A small amount of evidence has been pieced together indicating that carnitine may be active in the metabolism of fats or their derivatives". This conclusion was based primarily upon demonstrations around 1955 that pigeon liver contained an enzyme which catalyzed the formation of acetyl CoA from CoASH and carntine[2], and that carnitine addition increased rates of oxidation of palmitic acid by liver preparations[3]. With this relatively scant information, the reviewers[1] were bold enough to suggest that "..carnitine is strongly implicated in the utilization of fat in the animal body". Suppositions were greatly strengthened during the next few years by the acquisition of information on the kinetic properties and substrate specificities of purified carnitine acetyltransferase[4-6], and carnitine palmitoyltransferases[7-10]. In addition, the sites of control by carnitine of the rates of oxidation of long-chain fatty acids were tentatively identified[11-13]. In a review on carnitine actions in 1963 by Fritz[14], evidence was summarized in support of the hypothesis that carnitine acted to increase rates of fatty acid oxidation by mediating the translocation of fatty acyl groups across mitochondrial membranes to the sites of oxidation within mitochondria. This hypothesis, outlined in Figure 1, was based largely upon two sets of observations: (1) acyl groups of acyl CoA derivatives could not be oxidized by mitochondria in the absence of added carnitine, whereas acyl groups on acylcarnitine derivatives could readily be oxidized[11,12]; and (2) carnitine palmitoyltransferase reactions were reversible[7-10]. In the period since this hypothesis was initially proposed[11-14], results from many laboratories have validated sites of carnitine action on lipid metabolism (Figure 1). Consequently the major features illustrated in Figure 1 have gained general acceptance (for reviews, see References 15 and 16). In addition, information has emerged to include in the system an acylcarnitine translocate, a protein whose characteristics are different from those of known carnitine acyltransferases[17] (Figure 2). Carnitine palmitoyltransferases (CPT) possess unusually complex properties. Even though CPT has been isolated and characterized, and genes(s) now cloned, much remains to be learned concerning

1963 MODEL: CARNITINE AND MITOCHONDRIA

(FRITZ, ref. 14)

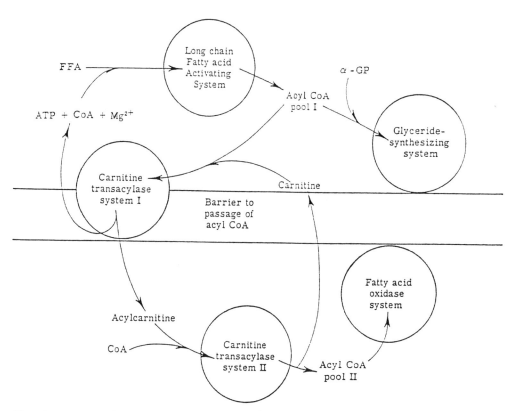

FIG. 1   Site of action of carnitine (**14**).

Explanation of symbols: Long-chain carnitine acyltranferase I is designated on the cytosolic and outer membrane site of the mitochondrial membrane which restricts the permeability of acyl CoA esters, and the carnitine acyltransferase II is designated on the matrix space side of the mitochondrial membrane in contact with the fatty acid oxidase system.

FFA: unesterified fatty acid; αGP:α-glycerophosphate the precise distribution of different isozymes on mitochondrial membranes, and the mechanisms by which CPT activities are regulated[16-25]. Lively controversies pertaining to these and related topics are much in evidence in contributions presented at this Symposium.

The physiological importance of the biochemical properties of carnitine and CPT has been dramatically demonstrated in patients with specific inborn errors in which carnitine uptake or metabolism is disturbed. Patients with muscle carnitine deficiency often manifest decreased rates of fatty acid oxidation (for reviews, see References 26-28). This may occur when there is an impairment in the transport of carnitine into muscle cells[26], or when carnitine acyltransferase activities are deficient[27] Theoretically, carnitine deficiency could also result from decreased rates of hydroxylation of butyrobetaine, or from increased renal excretion of carnitine. Some patients who suffer from severe carnitine deficiency are afflicted with muscle weakness, accompanied by an accumulation of lipids in specific types of muscle fibers[27]. Systemic carnitine deficiency and also secondary carnitine deficiency associated with organic acidemias often result in high ratios of acyl CoA to free CoASH concentrations in tissues[26,27]. The metabolic consequences of these changes are discussed below. The reader who is especially interested in the role of carnitine in clinical disorders of lipid metabolism in muscle, or in other tissues, is referred to recent reviews[27-29].

II. ACTIONS OF CARNITINE ON METABOLIC PROCESSES MEDIATED BY CARNITINE ACYLTRANSFERASES OTHER THAN CPT

Carnitine can serve as a substrate to accept acyl groups from a variety of acyl CoA derivatives, depending on the specificity of the carnitine acyltransferase. "Medium chain-length" carnitine acyltransferases (COT) rapidly catalyze the transfer of hexanoyl, octanoyl, and decanoyl groups[8,30]. Carnitine acyltransferases involved in the metabolism of branched-chain amino acids may generate compounds such as isovaleryl carnitine. Other carnitine acyltransferases are associated with the metabolism of fatty acids during chain shortening in peroxisomes (for reviews, see refs.15 and 16).

There must exist additional carnitine acyltransferases which possess different substrate specificities. For example, specific acylcarnitine derivatives appear in the urine of patients who suffer from various disorders of organic acid metabolism[31]. Administration of some drugs, such as valproic acid used for the treatment of epilepsy, also results in carnitine deficiency[32]. Urinary excretion of these types of carnitine derivatives results in the generation of secondary carnitine deficiency[29]. The possible physiological roles of such unique carnitine acyltransferases remain to be determined, However, these transferases could be of considerable significance in modulating levels of free CoASH in specific cell types. If CoASH levels were low because potentially toxic acyl CoA compounds had been formed, and sequestered in specific intracellular compartments, the

1989 MODEL: CARNITINE AND MITOCHONDRIA

(PANDE AND MURTHY, ref 17)

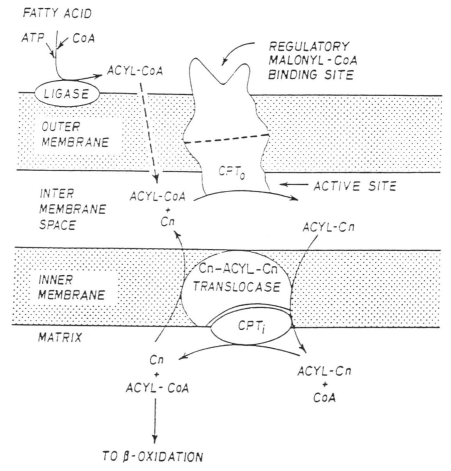

FIG. **2**. Schematic representation of the carnitine-dependent import of acyl groups into mitochondria. Being reversible, the same sequence allows also the export of acyl groups from the mitochondrial matrix. CPT_o, outer CPT; CPT_i, inner CPT; Cn, carnitine **(17)**.

addition of carnitine would liberate CoASH required for energy metabolism, and would facilitate the excretion of the toxic compound.

An "acute metabolic crisis" may occur in patients with systemic carnitine deficiency who are subjected to caloric deprivation, to exercise or to other stressful stimuli[26-28]. Symptoms associated with such "acute metabolic crises" are similar to those reported during attacks in Reye's syndrome. They include vomiting, stupor, confusion and coma in encephalopathic episodes. In addition, some patients with systemic carnitine deficiency are reported to manifest metabolic abnormalities such as increases in levels of plasma alanine or aspartate transaminase, and ammonium; decreases in plasma levels of glucose and prothrombin; metabolic acidosis; and increased urinary excretion of dicarboxylic acids and myoglobin[26-28]. Many of these abnormalities could be a metabolic consequence of decreased cellular levels of free CoASH, since most of the total CoA would be present primarily in various acyl esters. When tissue concentrations of free CoASH are sufficiently low, a demand for energy maintenance is likely to stimulate increased rates of reactions in alternative metabolic pathways. For example, an increased rate of omega oxidation of fatty acids would lead to increased formation of dicarboxylic acids. If administration of carnitine alleviated some of the metabolic symptoms associated with "acute metabolic crises", it might do so by liberating free CoASH from accumulated X-acyl CoA derivatives. This supposition is dependent upon the presence of specific carnitine acyltransferases capable of catalyzing the transfer of acyl groups from X-acyl CoA to carnitine.

The metabolic symptoms of "acute metabolic crisis", reported to occur in some patients with systemic carnitine deficiency, may therefore be a reflection of increased rates of reactions in "abnormal" compensatory metabolic pathways associated with an insufficiency of free CoASH. Carnitine administration would relieve the symptoms, in this scenario, simply by indirectly increasing levels of free CoASH in specific intracellular compartments.

III. ACTIONS OF CARNITINE MEDIATED BY MECHANISMS INDEPENDENT OF CARNITINE ACYLTRANSFERASES.

A. GENERAL CONSIDERATIONS

Carnitine acyltransfereases catalyze the transfer of acyl groups from acyl CoA derivatives to L(-)-carnitine but not to D(+)-carnitine. In fact, D(+)carnitine and acyl-D(+)-carnitine stereoisomiers are competitive inhibitors of these reactions[5-13] (for reviews, see References 14-16). It therefore follows that if D(+)- and L(-)-carnitine, or their derivatives, are equally effective in exerting influences on a system, the effects are not mediated by any of the characterized carnitine acyltransferases thus far investigated. It is possible, of course, that acyltransferases yet to be discovered could utilize D(+)-carnitine as a substrate.

Several laboratories[14-16] have shown that the apparent Km for carnitine in carnitine acyltransferase reactions (CAT, COT or CPT) is around 0.1 mM.

Product inhibition may occur at very high carnitine concentrations (unpublished observations). Levels of carnitine in epididymal luminal fluid reach concentration around 50 mM in epididymides of normal adult rats[33-36]. The presence of these elevated concentrations of carnitine would facilitate the formation of acetylcarnitine, which could then serve as an "energy reservoir" in special cases, such as that reported in flight muscles of blow flies[37]. A similar role for acetylcarnitine could possibly exist in the male reproductive tract, facilitating the generation in spermatozoa of acetyl CoA from acetylcarnitine in response to increased energy demands associated with sperm motility. By analogy with the metabolic controls reported in flight muscles of flies during their transition from rest to motion[35], there is a transition from rest to activity in motile spermatozoa after ejaculation. In both cases, increased activity would be accompanied by increases in rates of oxidation in the tricarboxylic acid cycle. High levels of acetylcarnitine could serve as an energy source, providing acetyl groups for citrate formation and oxidation, awaiting the generation of acetyl groups from other pathways. The very high levels of CAT in spermatozoa[33,34] are consistent with the possibility that acetylcarnitine could potentially function as an "acetyl group buffer" in spermatozoa of mammals, just as it appears to do in the flight muscles of insects[37].

Still, the extraordinarily high concentration of carnitine in epididymal luminal fluid (50 mM) suggests the possibility that carnitine may have functions other than serving as a substrate in the CAT reaction. Information to be presented in the following section indicates that L(-)- or D(+)-carnitine elicits the dispersion of spermatozoa, erythrocytes, and other types of cells, under conditions defined[38]. Perhaps, this property of carnitine may be utilized in vivo in preventing sperm aggregation within the epididymal lumen. In any case, these effects of carnitine on membranes of cells are not mediated by known carnitine acyltransfernases[38].

In addition to its influences on cell-cell interactions, carnitine exerts effects on several other membrane-related phenomena. For example, carnitine alters calcium ion channel activity[39]. It therefore appears probable that carnitine acts via mechanisms independent of any of the known carnitine acyltransferases by interacting with one or more components on cell membranes.

B.   ACTIONS OF CARNITINE ON CELL TO CELL INTERACTIONS

While investigating the actions of clusterin, a glycoprotein which elicits the aggregation of spermatozoa, erythrocytes and several other types of cells in suspension[40-42], we discovered that carnitine but not choline prevents the aggregation of several types of cells[38]. This phenomenon is not restricted to the inhibition of clusterin actions, since carnitine also blocks the effects of fetuin[38], or fibrinogen on the aggregation of erythrocytes (Fritz, Wong and Burdzy, submitted). D(+)- and L(-)-carnitine are equally effective in preventing the clustering of cells otherwise elicited by addition of either of these proteins. In the case of erythrocytes in suspension, addition of carnitine not only blocks but also reverses the aggregation. This phenomenon is thought to be mediated by processes independent of carnitine acyltransferases because identical effects can be obtained in glutaraldehyde-fixed erythrocytes by L(-)- or by D(+)-carnitine at equivalent concentrations[38]

The dispersion of red blood cells, or RBC ghosts, by carnitine is evident in cells preincubated with carnitine, washed extensively, and then incubated in the presence or absence of fibrinogen or clusterin (Fritz, Wong and Burdzy, submitted). These data demonstrate that carnitine bound to RBC membranes alters cell behaviour, either by decreasing attractive forces between the cells, or by increasing repulsive forces. Although D(+)-carnitine is as effective as L(-)-carnitine in preventing cell aggregation, other changes in the chemical structure of carnitine abolish its activity. Activity is lessened or eliminated by substituting an amino group for the trimethylammonium group; by removing the carboxyl group; or by removing the hydroxyl group from carnitine (Table 1)[38].

In considering possible mechanisms involved, we have investigated influences of sulfhydryl reagents on the binding of carnitine to RBC ghosts, and the correlations between binding of carnitine to effects on cell interactions. Experiments in progress support the possibility that sulfhydryl groups on erythrocyte membranes are required for the binding of carnitine and also for the actions of carnitine on the dispersion of cells. We are exploring the nature of the membrane-associated protein which bind carnitine, and which appear to be implicated in the modulation of cell interactions.

TABLE 1

Influences of Carnitine and Derivatives on the Inhibition of Aggregation of Glutaraldehyde-Fixed Human Erythrocytes Induced by Clusterin

L(-) or D(+) Carnitine or Derivative	ID_{50}^{*} (mM)
Carnitine	1.5
Decanoylcarnitine	0.5
Lauroylcarnitine	0.13
Myristoylcarnitine	0.05
Dimethylcarnitine ("Norcarnitine")	>12.5
Deoxycarnitine	>12.5
γ-Aminobutyrate	>25.0
Choline	>25.0
Acetyl ß-methylcholine	>25.0

*Approximate concentrations required for 50% inhibition of aggregation of cells incubated in the presence of clusterin (7.5 µg/ml)

*For other details, see Reference 38, from which this table is adapted.

C.   Actions of Carnitine on other Membrane-Related
     Phenomena

In recent years, influences of carnitine and its derivatives on myocardial tissue have become of increasing interest. During cardiac ischemia, and in some cardiac myopathies, long-chain acylcarnitine derivatives accumulate within specific membranes, and appear to modulate Ca^{2+} channel functions[39]. The actions of palmitoylcarnitine are reported to be similar in some respects to those of a pharmacological Ca^{2+} channel activator (Bay K8644) in K^+-depolarized smooth muscle preparations[39,43]. In this system, DL-palmitoylcarnitine was as effective as L(-)-palmitoylcarnitine, each having an EC_{50} of 75 µM[39].

During mild hypoxia, the number of α_1-adrenergic receptors in isolated canine myocytes increases two to three-fold, along with a corresponding increase in levels of long-chain acylcarnitine derivatives in the sarcolemma[44]. It is of considerable interest that inhibitors of carnitine palmitoyltransferase activity resulted not only in abolition of the increase in levels of palmitoylcarnitine, but also in an inhibition of the increase in numbers of α_1-adrenergic receptors which otherwise occurs in myocytes subjected to hypoxia[44]. From these and related data, Heathers et al.[44] concluded that levels of endogenous long-chain acylcarnitines in cardiac myocytes are directly correlated with the increase in α_1-adrenergic receptors induced by mild hypoxia.

In a different system, namely the Langendorff perfused rat heart, derivatives of long-chain DL-acylcarnitine esters have been reported to exert pronounced vasodilator activity, with a threshold concentration of 1 nM and an EC_{50} of 16 nM for the ethyl ester of palmitoylcarnitine[45]. In contrast, the administration of non-esterfied palmitoylcarnitine elicited coronary vessel vasoconstriction, accompanied by a cumulative irreversible depression of contractility[45]. The physiological significance of these intriguing observations remains unknown, but they clearly open a new sphere for investigations of pharmacological influences of carnitine derivatives. Since only esters of acylcarnitine derivatives were active, and since no differences in activity between D(+)- and L(-)-acylcarnitine derivatives were observed, it seems likely that the influences of ethyl esters of long-chain acylcarnitine derivatives on coronary blood vessels are mediated by mechanisms independent of known carnitine acyltransferases. Metabolic changes which occur during cardiac ischemia have been correlated with the accumulation of amphipathic molecules within cardiac membranes[46]. It is therefore of considerable interest to determine whether any derivatives of long-chain acylcarnitine esters are implicated in these processes.

Possible correlations between tissue and plasma levels of carnitine and myocardial ischemia or cardiac infarction, have been touched upon in a recent editorial on carnitine deficiency[47]. When administration of carnitine or specific acylcarnitine derivatives is followed by an alleviation of some of the symptoms of cardiac disease, it is conceivable that the effects observed could be mediated via interactions between carnitine or derivatives and specific components in membranes in muscle cells. This possibility is open to investigation.

IV.   SUMMARY AND CONCLUSIONS

Functions of carnitine have been reviewed extensively[14-16]. Well established functions include those involved in the controls of long-chain fatty acid oxidation in mitochondria (Figures 1 and 2); the shortening of fatty acid chains within peroxisomes; other metabolic pathways involving the oxidation of branched-chain a-keto acids; and the regulation of levels of CoASH to acyl CoA derivatives within specific intracellular compartments. It appears likely that many of the symptoms associated with systemic or secondary carnitine deficiency[26,27,28,47] are consequences of impairments in the above metabolic pathways, dependent upon carnitine. The important role of carnitine uptake in the etiology of systemic carnitine deficiency has recently been emphasized[48,49] along with an indication that carnitine administration may improve carnitine-responsive cardiomyopathy by increasing levels of carnitine uptake in myocytes[48]. In a consideration of the etiology and treatment of patients with carnitine deficiency, the delineation of carnitine transport and uptake mechanisms clearly is an important area for future research. In addition, it is crucial to define the role which carnitine may play in certain diseases of mitochondria in which muscle functions are impaired (for reviews, see References 27, 50 and 51).

In addition to the relatively well-defined functions of carnitine in intermediary metabolism listed above, other possible roles for carnitine appear to exist. Data reviewed in section III indicate several sytems in which interactions between carnitine and unidentified components in membranes are strongly implicated, independent of carnitine acyltransferases. It will be of considerable interest to determine mechanisms by which carnitine and its derivatives influence cell-cell interactions; Ca^{2+} channel functions; contractility of smooth muscles in blood vessels from different sites; and contractility of cardiac muscle.

ACKNOWLEDGMENTS

Researches reported from my laboratory were supported by grants from the Canadian Medical Research Council, and by Sigma Tau Pharmaceuticals, to whom I expess my sincere gratitude. I also thank Rosa Jara for typing this manuscript.

REFERENCES

1. Fraenkel, G. and Friedman, S., Carnitine, Vitamins and Hormones, 15, 73, 1957.

2. Friedman, S. and Fraekel, G., Reversible enzymatic acetylation of carnitine, Arch. Biochem. and Biophys., 59, 491, 1955.

3. Fritz, I.B., The effects of muscle extracts on the oxidation of palmitic acid by liver slices and homogenates, Acta Physiol. Skand., 34, 367, 1955.

4. Fritz, I.B., Schultz, S.K. and Srere, P.A., Properties of partially purified carnitine acetyltransferase, J. Biol. Chem., 238, 2509, 1963.

5. Fritz, I.B. and Schultz, S.K., Carnitine acetyltransferase II Inhibition by carnitine analogues and by sulfhydryl reagents, J. Biol. Chem. 240, 2188, 1965.

6. Chase, J.F.A., Pearson, D.J. and Tubbs, P.K., The preparation of crystalline carnitine acetyltransferase, Biochim. Biophys. Acta, 96, 162, 1965.

7. Norum, K.R., Palmityl-CoA: carnitine palmityltransferase. Purification from calf liver mitochondria and some properties of the enzyme, Biochim. Biophys. Acta 89, 95, 1964.

8. Norum, K.R., Palmityl-CoA: carnitine palmityltransferase. Demonstration of essential sulfhydryl groups on the enzyme, Biochim. Biophys. Acta 105, 506, 1965.

9. Kopec, B. and Fritz, I.B., Properties of a purified carnitine palmitoyltransferase, and evidence for the existence of other carnitine acyltransferases, Canad. J. Biochem., 49, 941, 1971.

10. Kopec, B. and Fritz, I.B., Comparison of properties of carnitine palmitoyltransferase I with those of carnitine palmitoyltransferase II, and preparation of antibodies, J. Biol. Chem., 248, 4069, 1973

11. Fritz, I.B. and Yue, K.T.N., Long-chain carnitine acyltransferase and the role of acylcarnitine derivatives in the catalytic increase of fatty acid oxidation induced by carnitine, J. Lipid Res., 4, 279, 1963.

12. Bremer, J., Carnitine in intermediary metabolism. The metabolism of fatty acid esters of carnitine by mitochondria, J. Biol. Chem., 237, 3628, 1962.

13. Fritz, I.B. and Marquis, N.R., The role of acylcarnitine esters and carnitine palmitoyltransferase in the transport of fatty acyl groups across mitochondrial membranes, Proc. Natl. Acad. Sci. USA, 54, 1226, 1965.

14. Fritz, I.B., Carnitine and its role in fatty acid metabolism, Adv. Lipid Res., 1, 285, 1963.

15. Bremer, J., Carnitine - metabolism and function. Physiol. Rev., 63, 1420, 1983.

16. Bieber, L.L., Carnitine, Ann. Rev. Biochem, 57, 261, 1988.

17. Pande, S.V. and Murthy, M.S.R., Carnitine: vitamin for an insect, vital for man, Canad. J. Biochem. and Cell Biol., 67, 671, 1989.

18. Bremer, J., The role of carnitine in intracellular metabolism, J. Clin. Chem. Clin. Biochem., 28, 297, 1990.

19. Murthy, M.S.R. and Pande, S.V., Malonyl-CoA binding site and the overt carnitine palmitoyltransferase activity reside on the opposite side of the outer mitochondrial membrane, Proc. Natl. Acad. Sci. USA, 84, 378, 1987.

20. Hoppel, C.L., Carnitine and carnitine palmitoyltransferase in fatty acid oxidation and ketosis, Fed. Proceed., 41, 2853, 1982.

21. Woeltje, K.F., Kuwajima, M., Foster, D.W. and McGarry, D., Characterization of the mitochondrial carnitine palmitoyltransferase enzyme system, J. Biol. Chem., 262, 9822, 1987.

22. Ramsay, R.R., The soluble carnitine palmitoyltransferase from bovine liver, Biochem. J., 249, 239, 1988.

23. Zammit, V.A., Corstorphine, C.G. and Kelliher, M.G., Evidence for distinct functional molecular sizes of carnitine palmitoyltransferases I and II in rat liver mitochondria, Biochem. J., 250, 415, 1988.

24. Brady, P.S., Dunker, H.K. and Brady, L.J., Characterization of hepatic carnitine palmitoyltransferase, Biochem. J., 241, 751, 1987.

25. Brady, L.J. and Brady, P.S., Regulation of carnitine palmitoyltransferase synthesis in spontaneously diabetic BB Wistar rats, Diabetes, 38, 65, 1989.

26. Engel, A.G. and Rebouche, C.J., Carnitine metabolism and inborn errors, J. Inher. Metab. Dis. 7, Suppl. 1, 38, 1984.

27. Engel, A.G., Carnitine deficiency syndromes and lipid storage myopathies, in <u>Myology Basic and Clinical</u>, Engel, A.G. and Banker, B.Q., Eds., McGraw-Hill, New York, 1663, 1986, Chap. 57.

28. DiMauro, S., Trevisan, C. and Hays, A., Disorders of lipid metabolism in muscle, Muscle and Nerve, 3, 369, 1980.

29. Feller, A.G. and Rudman, D., Role of carnitine in human nutrition, J. Nutr., 118, 541, 1988.

30. Solberg, H.E., Carnitine octanoyltransferase evidence for a new enzyme in mitochondria, FEBS. Lett., 12, 134, 1971.

31. Chalmers, R.A., Roe, C.R., Stacey, T.E. and Hoppel, C.L., Urinary excretion of l-carnitine and acylcarnitines by patients with disorders of organic acid metabolism: evidence for secondary insufficiency of l-carnitine, Pediatr. Res., 18, 1325, 1984.

32. Ohtani, Y., Endo, F. and Matsuda, I., Carnitine deficiency and hyperammonemia associated with valproic acid therapy, J. Pediatr., 101, 782, 1982.

33. Marquis, N.R. and Fritz, I.B., The distribution of carnitine, acetylcarnitine and carnitine acetyltransferase in rat tissues, J. Biol. Chem., 240, 2198, 1965.

34. Marquis, N.R. and Fritz, I.B., Effects of testosterone on the distribution of carnitine, acetylcarnitine and carnitine acetyltransferase in tissues of the reproductive system of the male rat, J. Biol. Chem., 240, 2197, 1965.

35. Brooks, D.E., Hamilton, D.W. and Mallek, A.H., Carnitine and glycerylphosphorylcholine in the reproductive tract of the male rate, J. Reprod. Fertil., 36, 141, 1974.

36. Hinton, B.T. and Setchell, B.P., Concentration and uptake of carnitine in the rat epididymis. A micropuncture study, in <u>Carnitine Biosynthesis, Metabolism and Functions</u>, Frenkel, R.A. and McGarry, J.D., Eds., Academic Press, New York, 237, 1980.

37. Childress, C.C., Sacktor, B. and Traynor, D.R., Function of carnitine in the fatty acid oxidase deficient insect flight muscle, J. Biol. Chem., 242, 754, 1967.

38. Fritz, I.B. and Burdzy, K., Novel Action of carnitine: inhibition of aggregation of dispersed cells elicited by clusterin in vitro, J. Cell. Physiol., 140, 18, 1989.

39. Spedding, M. and Mir, A.K., Direct activation of Ca^{2+} channel by palmitoylcarnitine, a putative endogenous ligand. Br. J. Pharmacol., 92, 457, 1987.

40. Fritz, I.B., Burdzy, K., Setchell, B. and Blaschuk, O., Ram rete testis contains a protein (clusterin) which influences cell-cell interactions in vitro, Biol. Reprod. 28, 1173-1983.

41. Blaschuk, O., Burdzy, K. and Fritz, I.B., Purification and characterization of a cell-aggregating factor (clusterin), the major glycoprotein in ram rete testis fluid, J. Biol. Chem., 258, 7714, 1983.

42. Fritz, I.B., Blaschuk, O. and Burdzy, K., Propeties of clusterin, a glycoprotein which elicits cell aggregation, and immunochemical determination of levels in ovine tissues, in Gonadal Proteins and Peptides, Sairam, M.R. and Atkinson, L.E., eds., World Scientific, Singapore, 312, 1984.

43. Patmore, L., Duncan, G.P. and Spedding, M., Interaction of palmitoylcarnitine with calcium antagonists in myocytes, Br. J. Pharmacol., 97, 443, 1989.

44. Heathers, G.P. Yamada, K.A., Kanter, E.M. and Corr, P.B., Long-chain acylcarnitines mediate the hypoxia-induced increase in α_1-adrenergic receptors on adult canine myocytes, Circulaion Res., 61, 735, 1987.

45. Criddle, D.N., Dewar, G.H., Wathey, W.B. and Woodward, B., The effects of novel vasodilator long chain acylcarnitine esters in the isolated perfused heart of the rat, Br. J. Pharmacol., 99, 477, 1990.

46. Corr, P.B., Gross, R.W. and Sobel, B.E., Amphipathic metabolites and membrane disfunctions in ischemic myocardium, Circ. Res., 55, 135, 1984.

47. Unsigned editorial, Carnitine deficiency, The Lancet, 335, 631, 1990.

48. Tien, I., DeVivo, C., Bierman, F., Pulver, P., De Meirleir, L.J., Cvitanovic-Sojat, L., Pagon, R.A., Bertini, E, Dionisi-Vici, C., Servidel, S. and DiMauro, S., Impaired skin fibroblast carnitine uptake in primary systemic carnitine deficiency manifested by childhood carnitine-responsive cardiomyopathy, Pediatr. Res., 28, 247, 1990.

49. Erikson, B.Q., Lindstedt, S. and Nordin, I., Transport of carnitine into cells in hereditary carnitine deficiency, J. Inher.Metab. Dis., 12, 108, 1989.

50. Shoffner, J.M. and Wallace, D.C., Oxidative phosphorylation diseases-disorders of two genomes, Adv. Genetics 19, 267, 1989.

DISCUSSION OF THE PAPER

G.A. COOK *(Memphis, TN)*: Does carnitine reverse clustering of cells or does it only prevent clustering?

FRITZ: Carnitine reverses and prevents the clustering of red blood cells elicited by fibrinogen or by clusterin. However, with several other cell types, such as TM 4 cells, carnitine prevents clustering elicited by clusterin, but does not reverse it.

P. BORUM *(Gainesville, FL)*: Do reticulocytes aggregate in the same manner as mature red cells? Is the sulfhydryl compound that is needed for the carnitine to function a small molecular weight compound or a large molecular weight compound?

FRITZ: We have not examined potential influences of carnitine on reticulocytes; however, the aggregation of lymphocytes elicited by clusterin is prevented by carnitine. We are in the process of attempting to identify components in red blood cell membranes which have sulfhydryl groups reactive with carnitine. Indirect evidence suggests that protein(s) are implicated, but we will not be sure until the active component(s) are isolated.

F. DI LISA *(Italy)*: Is D-Carnitine active?

FRITZ: L(-)-and D(+)-carnitine are equally active.

51. Zeviani, M., Bonilla, E., DeVivo, D.C. and DiMauro, S., Mitochondrial Diseaes, Neurologic Clinics 7, 123, 1989.

ABSTRACTS TO PART II

SKELETAL MUSCLE CARNITINE METABOLISM AS A MARKER OF EXERCISE PERFORMANCE IN DISEASE STATES. WR Hiatt and EP Brass. Departments of Medicine and Pharmacology, University of Colorado School of Medicine, Denver, Colorado 80262, and Case Western Reserve University, Cleveland, Ohio 44106.

Patients with chronic renal failure (CRF) on hemodialysis (HD), and patients with peripheral arterial disease (PAD) have a marked impaired in exercise performance. As carnitine is an important cofactor for muscle energy metabolism, changes in carnitine metabolism were evaluated in these diseases. In 8 patients on HD, peak exercise oxygen consumption (VO_2) was 16.2 ± 1.9 ml/kg/min (mean $\pm$ SD). The peak VO_2 in 10 patients with PAD was 14.8 ± 2.4, and in age-matched controls was 26.3 ± 3.5 ml/kg/min. In CRF, muscle (vastus lateralis) total carnitine content at rest was lower than controls (2320 ± 1190 vs. 3800 ± 940 nmol/g), and was correlated to peak exercise time ($r = 0.77$, $p < 0.05$); but the distribution between carnitine and acylcarnitines was normal. Patients on HD had a normal metabolic response to high intensity exercise as muscle short-chain acylcarnitine content increased from 130 ± 130 to 1380 ± 820 nmol/g ($p < 0.01$) which was correlated with the increase in muscle lactate content ($r = 0.88$, $p < 0.01$). In contrast, the patients with PAD had a normal muscle (gastrocnemius) total carnitine content at rest as compared to controls (2880 ± 510 vs. 3260 ± 810 nmol/g). The muscle short-chain acylcarnitine content at rest in PAD was inversely correlated with peak exercise time ($r = -0.70$, $p < 0.05$). In patients with PAD, the muscle short-chain acylcarnitine content increased from 420 ± 400 nmol/g at rest to 900 ± 630 nmol/g ($p < 0.05$) with exercise, which was correlated with exercise duration, and not to the lactate threshold. Thus, patients with CRF and PAD had a similar impairment in exercise performance. The metabolic basis for altered carnitine metabolism was different in the two diseases, as muscle in CRF was characterized by decreased total carnitine content, but a normal acylcarnitine distribution and response to exercise. Muscle in PAD was characterized by accumulation of short-chain acylcarnitines at rest, and at all levels of exercise. In both diseases, altered carnitine metabolism was correlated to exercise performance.

EFFECT OF L-THYROXIN (TX) AND HYPOPHYSECTOMY (HS) ON CARNITINE IN SERUM, LIVER, MUSCLE AND HEART OF RATS. E. Tombragel, C. McGraw, G. Hug and E. Landrigan. The Children's Hospital, Cincinnati, OH 45229.

Impaired carnitine (C) transport from serum (51μM) into muscle (4500 μM) may be treatable. Hypothyroid patients can have low C. Thus 4 of 6 normal (NL) rats, and 6 of 10 HS rats received i.m. 1.98 mcg/day (d) TX for 22 d. We compared total, free, short and long chain C, preTX and at d 3, 8, 23 in serum, at d 23 in liver, muscle, heart (nm/mg prot) in TX treated(+) vs untreated(-), in HS and in NL. Total and free C changed similarly at d 3, 8, 23. Representative values for total C (d 23): Serum C was lower in (-)HS '39.7 ± 3.5 vs (-)NL 87.4 ± 0.1, but it was normal in (+)HS 77.2 ± 5.8 vs (+)NL 85.5 ± 3.6. Liver C was lower in (-)HS 2.7 ± 0.5 vs (-)NL 5.5 ± 0.5, $p = 0.012$, but was normal in (+)HS 6.1 ± 0.9 vs (+)NL 5.1 ± 0.7, $p > 0.3$. Muscle C was lower in (-)HS 13.1 ± 1.1 vs (-)NL 21.5 ± 0.8, $p < 0.001$ but increased for (+)HS 19.0 ± 1.4 vs (+)NL 25.3 ± 1.3, $p = 0.016$. TX raised muscle C in (-)NL 21.5 ± 0.8 vs (+)NL 25.3 ± 1.3, $p = 0.032$. Heart C was not lower in (-)HS 19.3 ± 0.6 vs(-) NL 22.4 ± 1.6, $p = 0.2$ and TX produced a small increase in (-)HS 19.3 ± 0.6 vs (+)HS 21.3 ± 1.4, $p > 0.3$. Conclusion: TX raised C when it had been lowered by HS. TX raised C in NL muscle and insignificantly so in NL heart.

INTRAVENOUS FAT AND CARNITINE STATUS IN SURGICAL PATIENTS.
V. Tanphaichitr and P. Leelahagul. Division of Nutrition and Biochemical
Medicine, Department of Medicine and Research Center, Faculty of Medicine,
Ramathibodi Hospital, Mahidol University, Bangkok 10400, Thailand.

The effect of intravenous fat on carnitine status was assessed in 18 surgical
patients aged 33-72 years, with diseases of the gastrointestinal tract and on total
parenteral nutrition (TPN) for 14 days postoperatively. They were divided into
2 groups : fat-free and fat-supplemented groups. Both groups received a daily
supply of 2,700 kcal and 100 g of amino acids. The fat-supplemented group
received 50 g of fat 3 times a week. The results show that fat administration did
not lower their urinary and serum carnitine levels. This was due to the adequate
daily supply of 38.3 mmol of lysine and 25.5 mmol of methionine for carnitine
biosynthesis and the release of muscle carnitine induced by operation and fever.
The catabolic response of skeletal muscle in which most of the body carnitine
resides, were evidenced by the significant negative correlation between urinary
total carnitine and upper arm muscle circumference as well as significant positive
correlation between urinary total carnitine and nitrogen excretions. The effective
utilization of parenteral fat was supported by the non-significant difference in
plasma ketone body levels between these 2 groups.

Supplementary choline (CH) and pantothenate (PA) reduce serum and urinary
carnitine (CNE) concentrations in humans. Dodson, W.L. and Sachan, D.S. Dept.
Home Ec., Mississippi State, MS and Dept. Nutr. and Agri. Expt. Sta., Univ.of
Tenn., Knoxville.

We previously reported that supplementary CH/PA reduce serum and urinary CNE
concentrations in humans (FASEB J. 2, A1421, 1988). We have extended the number
of observations and have examined the effects of CH/PA on the serum and urinary
response to supplementary L-CNE. From day 0 to day 7 the control group (n=9)
took no supplements while the experimental group (n=20) took 13.5 mmol of CH and
1.4 mmol of PA daily. Additionally, both groups took 6.2 mmol of L-CNE on days
7 to 10. Whereas no changes occurred in the serum or urinary CNE concentrations
of the control group from day 0 to day 7, the experimental group had significant
decreases in serum and urinary nonesterified (NEC) and total CNE (TC) and in
urinary acid soluble acylcarnitine (ASAC). Supplementary L-CNE increased serum
CNE concentrations in both groups to approximately the same level. In contrast,
the renal NEC, ASAC, and TC clearances and excretion were significantly lower
(50%) in the experimental group than in the controls. When percent increase in
all fractions of urinary CNE excretion from day 7 to day 10 were examined, no
significance differences were found between the two groups; thus, conservation
of CNE beyond that needed to restore the serum concentration was not evident in
the experimental group. In conclusion, CH/PA significantly decreased serum and
urinary NEC and TC in humans, and supplementary L-CNE was needed to restore serum
concentrations.

PLASMA CREATINE KINASE (CK) LEVELS IN NORMAL AND HYPERAMMONEMIC MICE SHOWING MUSCLE CARNITINE DEPLETION CAUSED BY SODIUM BENZOATE (SB). I.A. Qureshi and A. Michalak. Research Center, Ste-Justine Hospital and University of Montreal, Montreal, Qc, Canada H3T 1C5.

We have reported earlier (FASEB J 1991;5:A593) that muscle carnitine reserves are depleted in normal and chronically hyperammonemic sparse-fur (spf) mutant mice, given either an acute treatment of SB (375 mg/kg b.w. i.p.) or a prolonged treatment (2% in drinking water). This reduction in total carnitine ranges from 44-58% in both normal and spf mice. We have now tested a hypothesis that the depletion of muscle carnitine could be due to an injury to muscle cells, by measuring plasma CK levels. The CK was significantly increased in normal mice given SB i.p. (416±80 U/l; mean±SEM, n=8, p<.01 Mann-Whitney's rank-sum test) compared to the control group given 0.9% NaCl i.p. (134±20 U/l, n=9); and in the prolonged SB treatment (373±34 U/l vs 61±11 U/l; p<.001). The untreated spf mice already showed a significant elevation of CK (432±30 U/l) and a depletion of muscle carnitine (62-65%), presumably caused by a chronic hyperammonemic state. Both the acute and prolonged SB treatment did not show any further significant change in CK levels in spf mice. The results indicate that both SB and hyperammonemia cause an injury to sarcoplasmic membrane, the exact site and mechanism of which remains to be determined. (Supported by MRC Canada #MA-9124).

L-CARNITINE PROTECTS FISH AGAINST ACUTE AMMONIA TOXICITY. G. Tremblay and T. Bradley, The University of Rhode Island, Kingston, RI 02881

Groups of 6-7 juvenile chinook salmon were injected ip with 0.25 M mannitol followed 1 hr later with ip ammonium acetate (AA). Graded signs of NH_3 toxicity were: 1, ataxia; 2, sinking but struggling to swim; 3, immobility; 4, convulsions; 5, death. At 10.75 mmol AA/kg body wt, 92% showed signs of NH_3 toxicity above grade 3 and 69% died (n=52). When L-carnitine (10-16 mmol/kg) was substituted for mannitol, 67% of the fish showed no signs of NH_3 toxicity and only 4% died (n=45). When trimethylamine oxide (TMAO, 10-16 mmol/kg) was substituted for mannitol, 44% of the fish showed no signs of NH_3 toxicity and only 18% showed signs above grade 2, with 16% dying (n-43). Substitution with acetylcarnitine, betaine, or choline at 10-12 mmol/kg killed the fish before the challenge with AA could be administered.

Since salmonids lack a urea cycle, accelerated ureagenesis cannot account for the observed protection, as described for mice (1). Protection by TMAO suggests a mechanism unrelated to the role of carnitine in fatty acid oxidation.

---

1. Costell, M et al., (1984) Biochem. Biophys. Res. Commun. 120:726-733.

CARNITINE IN THE TREATMENT OF OBESE SUBJECTS.

G.E.Narváez Pérez, J.J.Alvarez Casado, A.W.Caprile, J.C.Caubet and M.Mollerach.

Laboratorio de Evaluaciones Morfofuncionales (LABEMORF) y CASASCO S.A.I.C.,Amenabar 783 (1426). Buenos Aires – Argentina.

The anthropometric and functional variations were studied in obese subjects (O.S.) and control group (C.G.) that were submitted: 1- to general training program (G.T.P.) of physical adaptation of 9 weeks; and 2- specific training program (S.T.P.) of intermittent physical work near 100 % of the VO2 Max.in 15 sec. period alternating with 15 sec. of rest during 25-40 min. three times a week, over a period of 16 weeks. Moreover,the O.S. received an oral solution of 3 g per day of L-Carnitine by periods of 4 weeks and 2 weeks of suspension (1).

After G.T.P. significant increase of VO2 and tolerance to more intense work load in both groups were observed.How ever,the anthropo metric changes in fat and muscle weight,only were significant in the O.S.carnitine -loaded,after the S.T.P.,see fig.

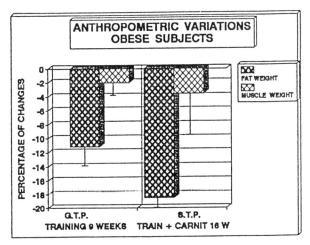

These data suggested, that FFA mobilization during intermittent work (15 x 15 sec.)observed in trained subjects subject (2) and the increased utilization by muscle in carnitine -loaded subjects, (3)should explain the significant decrease of fat weight,but without significant changes in muscle weight in the O.S. However,is necessary a previous physical adaptation for carry out the 15 x 15 intermittent work.

(1) Narváez et al.Annals of ACSM 36th Annual Meeting, 1989.
(2) Bigita Esssen,Acta Physiol.Scand.,1980.
(3) Narváez et al.Pergamon Press.pp 44-45,1986.

SKELETAL MUSCLE CARNITINE METABOLISM AS A MARKER OF EXERCISE PERFORMANCE IN DISEASE STATES. WR Hiatt and EP Brass. Departments of Medicine and Pharmacology, University of Colorado School of Medicine, Denver, Colorado 80262, and Case Western Reserve University, Cleveland, Ohio 44106.

Patients with chronic renal failure (CRF) on hemodialysis (HD), and patients with peripheral arterial disease (PAD) have a marked impaired in exercise performance. As carnitine is an important cofactor for muscle energy metabolism, changes in carnitine metabolism were evaluated in these diseases. In 8 patients on HD, peak exercise oxygen consumption (VO_2) was 16.2 ± 1.9 ml/kg/min (mean $\pm$ SD). The peak VO_2 in 10 patients with PAD was 14.8 ± 2.4, and in age-matched controls was 26.3 ± 3.5 ml/kg/min. In CRF, muscle (vastus lateralis) total carnitine content at rest was lower than controls (2320 ± 1190 vs. 3800 ± 940 nmol/g), and was correlated to peak exercise time ($r = 0.77$, $p < 0.05$); but the distribution between carnitine and acylcarnitines was normal. Patients on HD had a normal metabolic response to high intensity exercise as muscle short-chain acylcarnitine content increased from 130 ± 130 to 1380 ± 820 nmol/g ($p < 0.01$) which was correlated with the increase in muscle lactate content ($r = 0.88$, $p < 0.01$). In contrast, the patients with PAD had a normal muscle (gastrocnemius) total carnitine content at rest as compared to controls (2880 ± 510 vs. 3260 ± 810 nmol/g). The muscle short-chain acylcarnitine content at rest in PAD was inversely correlated with peak exercise time ($r = -0.70$, $p < 0.05$). In patients with PAD, the muscle short-chain acylcarnitine content increased from 420 ± 400 nmol/g at rest to 900 ± 630 nmol/g ($p < 0.05$) with exercise, which was correlated with exercise duration, and not to the lactate threshold. Thus, patients with CRF and PAD had a similar impairment in exercise performance. The metabolic basis for altered carnitine metabolism was different in the two diseases, as muscle in CRF was characterized by decreased total carnitine content, but a normal acylcarnitine distribution and response to exercise. Muscle in PAD was characterized by accumulation of short-chain acylcarnitines at rest, and at all levels of exercise. In both diseases, altered carnitine metabolism was correlated to exercise performance.

CHANGES IN MITOCHONDRIAL ACTIVITY CAUSED BY AMMONIUM SALTS AND PROTECTIVE EFFECT OF CARNITINE. M. Bellei, D. Battelli, E. Arrigoni-Martelli[*], U. Muscatello and V. Bobyleva. Institute of General Pathology, University of Modena, 41100, Italy and [*]Sigma Tau, Rome, Italy.

Previous findings have shown that ammonium acetate treatment in vivo adversely affects the functioning of liver mitochondria, and that L-carnitine administration partially prevents the inhibition of mitochondrial respiration due to ammonium.

In order to get better insight into the mechanism of this intoxication and the specificity of prevention by carnitine, isolated rat liver mitochondria were treated in vitro with ammonium acetate and related compounds, and the effect of carnitine was investigated in these experimental models. It was found that ammonium salts, added to isolated rat liver mitochondria, deviate α-ketoglutarate to glutamate synthesis, thus decreasing its availability as respiratory substrate. As a consequence a decrease of respiratory rate is observed which is paralleled by progressive mitochondrial swelling. It was demonstrated that L-carnitine may abolish this swelling thus improving structural and metabolic state of mitochondria.

PART III

CARNITINE PALMITOYL TRANSFERASES

INTRODUCTION TO CPT

L. L. Bieber
Department of Biochemistry
Michigan State University
East Lansing, MI 48824

The purpose of this introduction is to briefly review the history of CPT to set the stage for today's discussion about CPT. One could designate this topic as "too many CPTs?". For this discussion, carnitine palmitoyltransferase (CPT) will refer to the mitochondrial enzyme or enzymes that catalyze the reversible transfer of long-chain acyl-CoAs to long-chain acylcarnitines. Since palmitoyl-CoA is the substrate most frequently used to study this reaction, the broad acyl-CoA specificity enzyme was referred to as CPT.

In 1971, Fritz and colleagues described the purification of CPT from calf liver, and in 1973, they reported the interesting observation that CPT-I could be converted to a different form of the enzyme, CPT-II, by treatment with urea. Almost two decades and numerous publications later, the basis for this intriguing observation has not been elucidated. If anything, the field has become more confusing, resulting in different designations for CPT, such as CPT_A/CPT_B and "overt" and "latent" CPT. The different names arose because the preparations under investigation seemed to be different forms of the same enzyme activity. For example, during the late 1970s and early 1980s, considerable effort was devoted to investigating easily soluble and tightly membrane-bound enzymes that exhibited CPT activity in liver. Some clarification began to occur with the recognition that peroxisomes and microsomes also contain medium-chain and long-chain carnitine acyltransferase activity. These enzymes, like CPT, are broad acyl-CoA-specific enzymes that also use both medium-chain and long-chain acyl-CoAs as substrates. Because COTs can use palmitoyl-CoA, they often are referred to as CPTs, but when standard biochemical criteria, i.e. $V_{max}/K_{0.5}$ for 10-carbon versus 16-carbon acyl-CoAs, are applied, they exhibit a preference for the 10-carbon chain-length substrates. Thus, kinetic criteria indicate the preferred substrates are medium-chain, not long-chain, acyl-CoAs. A major difference between the mitochondrial CPT and the COTs is that the microsomal and peroxisomal enzymes exhibit Michaelis-Menten kinetics; mitochondrial CPT does not. The ability to use medium-chain acyl-CoA led to their designation as carnitine octanoyltransferases (COT). Like CPT, octanoyl was designated because octanoyl-CoA was initially used to study this activity. Although microsomal and peroxisomal COT exhibit medium- and long-chain trasnferase activity, their function is not as clear as is the function of mitochondrial CPT. The peroxisomal enzyme seems to function as medium-chain acyltransferase to shuttle chain-shortened β-oxidation products out of peroxisomes as acylcarnitines. Rigorous characterization of "easily solubilized" CPT of liver mentioned previously showed it to be of peroxisomal origin with the substrate specificity of a COT and a primary amino acid sequence showing < 30% homology with the primary amino acid sequence of CPT from rat liver.

The use of <u>different assumptions</u> to interpret data has contributed to the ambiguity in the field. For example, two general models have been used to interpret much of the mitochondrial CPT data. In one model (model 2), two different catalytic proteins are envisioned, i.e. different enzymes such as CPT_A and CPT_B, with only one being sensitive to malonyl-CoA. In the other model (model 1), regulation of CPT, i.e. inhibition by malonyl-CoA, is thought to be due to association of a regulatory subunit with a catalytic protein. Thus, the different forms or expressions of CPT activity could be due, in part, to the presence or absence of a regulator subunit interacting with a single catalytic species of CPT or due to two different catalytic species. In this regard, the original CPT-I and CPT-II model, in which CPT-I was converted to CPT-II by urea, and the CPT_o and CPT_i models are compatible with a single catalytic CPT in mitochondria. In contrast, the CPT_A/CPT_B, "overt" and "latent," as well as the easily soluble/tight membrane association models all require the existence of at least two different catalytic proteins. Regardless, most, if not all, investigators in the field agree that CPT exists in at least two compartments in mitochondria.

The recent demonstration that addition of extracts, from liver mitochondria containing the malonyl-CoA binding protein devoid of CPT, to preparations of CPT that are insensitive to malonyl-CoA can restore malonyl-CoA sensitivity, has important implications about regulation of CPT and about the relationship of the different forms of CPT to each other. The table summarizes some of the enzymological data showing similarities and differences between COTs and mitochondrial CPT. In this table, COT = the enzymes associated with microsomes and peroxisomes that exhibit medium-chain and long-chain carnitine acyltransferase activity, but have a kinetic preference for medium-chain acyl-CoAs, i.e. $V_{max}/K_{0.5}$ for medium-chain acyl-CoA is > the $V_{max}/K_{0.5}$ for long-chain acyl-CoAs.

The following conclusions can be made from the data summarized in Table 1:

1. Mitochondrial, microsomal, and peroxisomal medium-chain/long-chain carnitine acyltransferases are antigenically distinct.
2. Mitochondrial, microsomal, and peroxisomal medium-chain/long-chain carnitine acyltransferases are different kinetically.
3. An octylglucoside soluble β-oxidation/malonyl-CoA sensitive CPT complex can be isolated from heart mitochondria.
4. Malonyl-CoA sensitivity can be restored to purified CPT.
5. The COT/CPT activity of "intact" microsomes, but not mitochondria, is inhibited by palmitoyl-CoA.
6. Membrane-bound microsomal and mitochondrial COT/CPT can be inhibited by malonyl-CoA with similar K_Is.

7. Etomoxiryl-CoA and aminocarnitine both inhibit microsomal, peroxisomal and mitochondrial COT/CPT activity, but the inhibition patterns for the individual enzyme systems are different.

8. $\dfrac{V_{max}\ C10/K_0.5\ (C10)}{V_{max}\ C10/K_0.5\ (C10)}$ are different for the mitochondrial, microsomal and peroxisomal COT/CPT enzymes.

9. The primary structure of mitochondrial CPT is very different ($< 30\%$ homology) than peroxisomal COT.

10. OMV exhibit characteristics of both the mitochondrial and the microsomal COT/CPT systems.

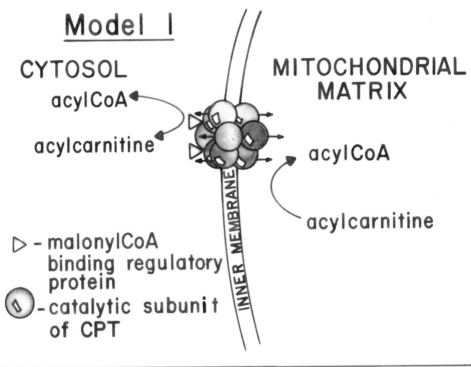

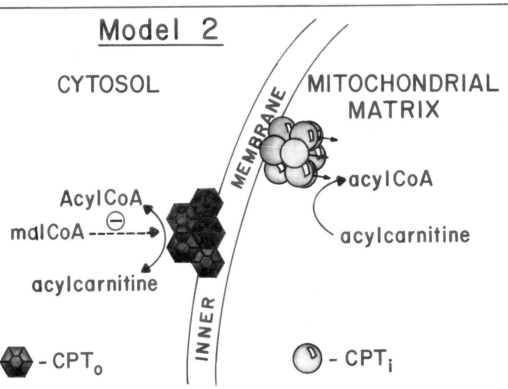

TABLE 1
Properties of Mitochondrial CPT, Peroxisomal COT, and Microsomal COT

Characteristic	Mitochondrial CPT$_o$,[a] CPT$_i$,[b]	Peroxisomal COT	Microsomal COT[c]
Membrane-Bound	Yes[6]	No; located in matrix[6,8]	Yes;[8,13] located on outer surface of ER[25]
Kinetic Characteristics	Heart complex, allosteric cooperative substrate binding with acyl-CoA and L-carnitine[27]; Liver CPT is oligomeric[1,18]	Michaelis-Menten Hill n = 1-1.2 for acyl-CoA mouse liver enzyme[9,10]	Michaelis-Menten Hill n ~ 1 for acyl-CoA and L-carnitine[14]
Antigenic Characteristics: A. Immunoprecipitation by anti-mitochondrial CPT	Yes[2]	No[2]	No
B. Immunoprecipitation by anti-peroxisomal COT	No[14]	Yes[10]	No
Palmitoyl-CoA Inhibition	a. OMV CPT[d] inhibited by > 18 μM[3] b. IMV CPT[e] not inhibited by 100 μM[3]	Not inhibited by 300 μM[11]	Membrane-bound I$_{50}$ = 11 μM[14,26]
Malonyl-CoA Inhibition	a. CPT$_o$ K$_i$ 1-2 μM[4] b. CPT$_i$ not inhibited[4]	Soluble enzyme not inhibited,[1] but membrane-bound enzyme in peroxisomes inhibited with I$_{50}$ = 2.2 μM[12]	Membrane-bound I$_{50}$ = 5 μM[14,26]
Etomoxiryl-CoA Inhibition	a. CPT$_o$ I$_{50}$ = 3 nM[5] b. Purified (beef heart) CPT decreases Hill n[16] c. OMV K$_i$ > 100 nM[15] [f]	K$_i$ = 1 μM, mixed type inhibitor[16]	a. Membrane-bound I$_{50}$ = 0.7 μM[16] b. CHAPS solubilized enzyme I$_{50}$ = 1.5 μM[16]
L-Aminocarnitine Inhibition	a. OMV CPT[d] I$_{50}$ = 250 μM[7] b. IMV CPT[e] I$_{50}$ = 25 μM[7] c. Purified CPT I$_{50}$ - 25 μM[7]	Not inhibited by 2 mM (data not published)	Membrane-bound I$_{50}$ = 500 μM for DL-aminocarnitine[14]
$\frac{V_{max} C10/K_{0.5} (C10)}{V_{max} C16/K_{0.5} (C16)}$	Heart mitochondria $\cong$ 4, but log K$_{0.5}$ L-carnitine vs. acyl chain length	~ 10	~ 30
Primary Structure	Amino acid sequence; CPT$_i$ of rat liver[17] and human liver have considerable homology[18,20] but < 30% homology with peroxisomal enzyme	rat liver sequence known[19]	Not determined
Restoration of Malonyl-CoA Inhibition	Addition of 86,000 dalton malonyl-CoA binding proteins results in ~ 50% inhibitability by malonyl-CoA for liver CPT$_i$[21-23] and for purified CPT$_i$ from heart[24]	Not determined	Not determined

[a] Malonyl-CoA sensitive form of CPT in contact with the cytosol
[b] Form of CPT in contact with the matrix of mitochondria
[c] Copies of submitted data available on request
[d] CPT activity of outer membrane-enriched vesicles
[e] CPT activity of inner membrane-enriched vesicles
[f] Estimated from data in Table I of reference 15

REFERENCES

1. Miyazawa, S., Ozasa, H., Osumi, T., and Hashimoto, T., Purification and properties of carnitine octanoyltransferase and carnitine palmitoyltransferase from rat liver, J. Biochem., 94, 529, 1983.

2. Kerner, J., and Bieber, L., Isolation of malonyl-CoA sensitive CPT-β-oxidation enzyme complex from heart mitochondria, Biochemistry, 29, 4326, 1990.

3. Murthy, M.S.R., and Pande, S.V., Some differences in the property of CPTase activities of mitochondrial outer and inner membranes, Biochem. J., 248, 727, 1987.

4. McGarry, J.D., Leatherman, G.F., and Foster, D.W., CPTase I: The site of inhibition of fatty acid oxidation by malonyl-CoA, J. Biol. Chem., 253, 4128, 1978.

5. Declercq, P.E., Falck, J.R., Kuwajima, M., Tyminski, H., Foster, D.W., and McGarry, J.D., Characterization of the mitochondrial carnitine palmitoyltransferase enzyme system: Use of inhibitors, J. Biol. Chem., 262, 9812, 1987.

6. Healy, M.J., Kerner, J., and Bieber, L.L., Is overt CPT of liver peroxisomal COT?, Biochem. J., 249, 231, 1988.

7. Murthy, M.S.R., Ramsay, R.R., and Pande, S.V., Acyl-CoA chain length affects the specificity of various carnitine palmitoyltransferases with respect to carnitine analogs, Biochem. J., 267, 273, 1990.

8. Markwell, M.A.K., and Bieber, L.L., Localization and solubilization of a microsomal carnitine acetyltransferase in rat liver, Arch. Biochem. Biophys., 172, 502, 1976.

9. Bieber, L.L., and Farrell, S., Carnitine acyltransferases, in *The Enzymes* 5, 1983, 627.

10. Farrell, S.O., Fiol, C.J., Reddy, J.K., and Bieber, L.L., Properties of purified carnitine acyltransferases of mouse liver peroxisomes, J. Biol. Chem., 259, 13089, 1984.

11. Farrell, S.O., and Bieber, L.L., Carnitine octanoyltransferase of mouse liver peroxisomes: Properties and effect of hypolipidemic drugs, Arch. Biochem. Biophys., 222, 123, 1983.

12. Derrick, J.P., and Ramsay, R.R., L-Carnitine acyltransferase in intact peroxisomes is inhibited by malonyl-CoA, Biochem. J., 262, 801, 1989.

13. Markwell, M.A.K., Tolbert, N.E., and Bieber, L.L., Comparison of the carnitine, Arch. Biochem. Biophys., 176, 479, 1976.

14. Lilly, K., and Bieber, L., Kinetic characterization and DL-aminocarnitine inhibition of rat liver carnitine octanoyltransferase, Life Science Advances, in press, 1991.

15. Murthy, M.S.R., and Pande, S.V., Characterization of a solubilized malonyl-CoA-sensitive CPTase from mitochondrial outer membrane as a protein distinct from the malonyl-CoA-insensitive CPTase of the inner membrane, Biochem. J., 268, 599, 1990.

16. Chang, C., Lilly, K., Kerner, J., Van Renterghem, R., and Bieber, L.L., The effect of etomoxiryl-CoA on different carnitine acyltransferases, submitted for publication, 1991.

17. Woeltje, K.F., Esser, V., Weis, B.C., Sen, A., Cox, W.F., McPhaul, M.J., Slaughter, C.A., Foster, D.W., and McGarry, J.D., Cloning, sequence and expression of cDNA encoding rat liver mitochondrial CPT-II, J. Biol. Chem., 265, 10720, 1990.

18. Finocchiaro, G., Colombo, I., and DiDonato, S., Purification, characterization and partial amino acid sequences of carnitine palmitoyl-transferase from human liver, FEBS Lett., 274, 163, 1990.

19. Chatterjee, B., Song, C.S., Kim, J.M., and Roy, A.K., Cloning, sequencing and regulation of rat liver carnitine octanoyltransferase: Transcriptional stimulation of the enzyme during peroxisome proliferation, Biochemistry, 27, 9000, 1988.

20. Finocchiaro, G., Taroni, F., Rocchi, M., Martin, A.L., Colombo, I., Tarelli, G.T., and DiDonato, S., cDNA cloning, sequence analysis and chromosomal localization of the gene for human carnitine palmitoyltransferase, Proc. Natl. Acad. Sci., 88, 661, 1991.

21. Woldegiorgis, G., Fibich, B., Contreras, L., and Shrago, E., Reconstitution of a purified malonyl-CoA sensitive carnitine palmitoyltransferase from rat liver mitochondria, FASEB J., 4, #7, Paper #3106, 1990.

22. Ghadiminejad, I., and Saggerson, E.D., Carnitine palmitoyltransferase (CPT$_2$) from liver mitochondrial inner membrane become inhibitable by malonyl-CoA is reconstituted with outer membrane malonyl-CoA binding protein, FEBS Lett., 269, 404, 1990.

23. Ghadiminejad, I., and Saggerson, E.D., The relationship of rat liver overt carnitine palmitoyltransferase to mitochondrial malonyl-CoA binding entity and to the latent palmitoyltransferase, Biochem. J., 270, 787, 1990.

24. Chung, C., Woldegiorgis, G., and Bieber, L., Restoration of malonyl-CoA sensitivity to purified rat heart mitochondrial carnitine palmitoyltransferase by addition of protein fraction(s) from an 86 kD malonyl-CoA binding immunoaffinity column, FASEB. J., 5, Abstract, 1991.

25. Valkner, K., and Bieber, L.L., The sidedness of carnitine acetyltransferase and carnitine octanyltransferase of rat liver endoplasmic reticulum, Biochim. Biophys. Acta., 689, 73, 1982.

26. Lilly, K., Bugaisky, G.E., Umeda, P.K., and Bieber, L.L., The medium-chain carnitine acyltransferase activity associated with rat liver microsomes is malonyl-CoA sensitive, Arch. Biochem. Biophys., 280, 167, 1990.

27. Fiol, C.J., Kerner, J., and Bieber, L.L., The effect of malonyl-CoA on the kinetics and substrate cooperativity of membrane-bound carnitine palmitoyltransferase of rat heart mitochondria, Biochim. Biophys. Acta, 916, 482, 1987.

BIOCHEMICAL, MOLECULAR BIOLOGICAL AND TOPOGRAPHICAL FEATURES OF THE MITOCHONDRIAL CARNITINE PALMITOYLTRANSFERASE SYSTEM

J. Denis McGarry[¶][§], Anjan Sen[¶], Nicholas F. Brown[¶],
Victoria Esser[¶], Brian C. Weis[¶], and Daniel' W. Foster[¶]

Departments of Internal Medicine[¶] and Biochemistry[§] and the Center for Diabetes Research, University of Texas Southwestern Medical Center, Dallas, Texas 75235

I. INTRODUCTION

First conceptualized as a mechanism for mitochondrial fatty acid transport some thirty years ago,[1,2] the carnitine palmitoyltransferase (CPT) system has come to occupy a prominent position in current discourses on fuel homeostasis. The present occasion is a case in point. Renewed interest in this area has been prompted by three major developments over the past decade or so: (i) recognition that CPT I plays a pivotal role in the regulation of fatty acid oxidation by virtue of its potent inhibition by malonyl-CoA;[3] (ii) the growing interest in this locus as a potential site for pharmacological intervention in the treatment of diabetes mellitus;[4,5] and (iii) the rapidly expanding list of individuals with documented inheritance of CPT deficiency.[6,7] It is not our goal here to attempt a comprehensive review of the subject, as some of its historical aspects will no doubt have been dealt with already by Dr. Irving Fritz, a pioneer in the field. Rather, we shall focus on recent efforts aimed at elucidating the CPT system in terms of its biochemical, molecular biological and topographical properties.

II. BIOCHEMICAL AND PHYSICAL PROPERTIES OF THE CPT ENZYMES

The mitochondrial transport of long chain fatty acids, as presently understood, is depicted in Figure 1. Although in a functional sense the model has not changed since its original formulation, two important details have come to light from the work of Pande and colleagues. One is the presence of a carnitine-acylcarnitine translocase in the inner membrane;[8,a] the other is reassignment of CPT I from the inner to the

---

[a]Relative to the CPT enzymes, the translocase has not been studied in great detail and will not be considered further here.

FIGURE 1.

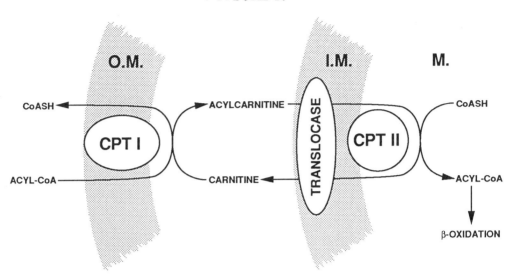

Mitochondrial transport of long chain fatty acids.
CPT, carnitine palmitoyltransferase; OM, outer membrane; IM, inner-membrane; M, matrix.

outer membrane.[9] Despite intensive efforts, dissection of the system into its component parts has proved to be a formidable task and one that has generated controversy. There is, however, general agreement on two points. First, while membrane-bound, CPT I, but not CPT II, is susceptible to inhibition by malonyl-CoA (reversible) and agents of the TG-CoA[b] class (irreversible). Second, detergent solubilized preparations of CPT are insensitive to these agents. What is still debated is whether in a given tissue CPT I and CPT II represent the same or different proteins, and how inhibitors interact with CPT I, i.e., with the polypeptide itself or with an associated regulatory protein.[3,10-15]

Our approach to these questions included (a) competitive binding studies using intact mitochondria exposed to mixtures of malonyl-CoA, TG-CoA and 2-bromopalmitoyl-CoA (BP-CoA) plus carnitine;[16,17] (b) treatment of mitochondria from various sources with different detergents followed by analysis of the solubilized and residual fractions for malonyl-CoA/TG-CoA sensitive and insensitive forms of CPT;[18,19] (c) exposure of intact mitochondria to [³H]TG-CoA with subsequent sizing of the [³H]TG-CoA binding protein and its separation from CPT catalytic activity by ion exchange chromatography;[16,17,19] and

---

[b]The abbreviations used are: TG-CoA tetradecylglycidyl-CoA; etomoxir-CoA, 2 - [6- (4 - chlorophenoxy) hexyl] oxirane -2- acyl-CoA; DEHP, diethylhexylphthalate; SDS, sodium dodecylsulphate; PAGE, polyacrylamide gel electrophoresis; PVDF, polyvinylidene diflouride.

(d) the use of a polyclonal antibody raised against purified, active CPT from rat liver to probe Western blots of mitochondrial membranes before and after their extraction with Tween-20.[18,19] The main conclusions were as follows:

(i) All of the CPT I inhibitors tested bound to intact muscle mitochondria at a single, common locus. Since one of these agents (BP-CoA) almost certainly acts at the catalytic center of CPT I it seems likely that malonyl-CoA and TG-CoA must also interact at this site, rather than with a regulatory protein.

(ii) In any given tissue CPT I and CPT II are distinct proteins, the former being of larger monomeric size.

(iii) Within a species CPT II is probably the same protein body wide, whereas CPT I exists as tissue specific isoforms.

(iv) Rat CPT II is immunologically similar to, but migrates on SDS gels slightly more slowly than, its counterpart in the mouse, monkey and human.

(v) Regardless of source, CPT I, but not CPT II, is inactivated by strong detergents.

(vi) The active CPT isoform (with apparent M_r of ~70,000) that has been purified from various rat and human tissues is CPT II; CPT I has yet to be isolated in pure form from any source.

We concede that the above viewpoints require further validation before they can be expected to gain general acceptance. At the present time, however, we consider them to be consistent with the weight of evidence. Accordingly, they form the basis of subsequent discussion.

III. MOLECULAR BIOLOGICAL ASPECTS OF CPT

A. RAT CPT II

It seemed that a deeper understanding of the CPT system would be greatly facilitated if we had knowledge of the primary structures of the two isozymes involved. This required a molecular biological approach and we chose to begin with rat liver CPT II since we already had the purified enzyme as well as an antibody against it. Using classical techniques a full length cDNA clone was ultimately obtained and sequenced.[20] It predicted a precursor protein of 658 amino acids containing a mitochondrial leader sequence of undetermined length (see below). Identity of the cDNA clone was confirmed in a number of ways, the most convincing being that after transfection into COS cells, the cDNA elicited a selective and marked induction of CPT II activity in the isolated mitochondria.[20] Computer analysis of the predicted amino acid sequence established that the protein is not particularly hydrophobic and that it probably forms a relatively loose association with the mitochondrial inner membrane.

There has been controversy surrounding the structure of the N-terminal region of mature rat CPT II. In our hands, purified preparations of the rat liver enzyme have been found to be blocked at the N-terminus, precluding direct amino acid analysis of this region. The predicted sequence from our cDNA clone suggested that the nascent protein contains a mitochondrial leader peptide of 20-30 residues, but, as expected, provided no information on the site of its cleavage. The Brady laboratory,[21] also working with purified CPT from rat liver (presumably CPT II), reported an octapeptide sequence which they had obtained after unblocking the N-terminus of the mature protein by chemical means. This stretch of amino acids was not represented in our cDNA clone. Based on analysis of a rat genomic library, Brady et al[22] suggested that our cDNA lacked a 5' segment that encodes their octapeptide as well as the true leader sequence of the precursor molecule. In an attempt to resolve the issue the following experiment was carried out. We transcribed our full length cDNA and used the product for in vitro translation in the presence of ^{35}S-methionine, as described previously.[20] The mixture was then incubated with rat liver mitochondria after which non-imported material was destroyed by trypsin. Detergent treatment of the mitochondria, followed by immunaffinity chromatography of the extract (using a covalently linked system of agarose-protein A- anti CPT II antibody), resulted in the rapid isolation of a single, labeled protein. On SDS gels this material proved to be slightly smaller than the initial translation product, and comigrated with pure CPT II. After electrophoretic transfer to a PVDF membrane the protein was subjected to automated Edman degradation and the cleaved N-terminal residue of each cycle was taken for scintillation counting. The first ^{35}S-methionine was released on the 18th cycle. Based on the deduced amino acid sequence from our cDNA clone, this result indicated that the N-terminus of the mature protein must have been the serine-26 residue of the initial translation product. Further support for this conclusion comes from the finding that precisely the same cleavage site (i.e., between Leu-25 and Ser-26) has recently been established for the leader peptide of human liver CPT II[23] (see below). We have no explanation for the findings reported by Brady et al.[21,22]

Northern analysis of poly A[+] RNA from rat liver using the labeled cDNA as a probe revealed a single species of mRNA of ~2.5kb in size.[24] Message intensity increased dramatically in livers from animals fed a diet containing DEHP, a known inducer of peroxisomal and mitochondrial carnitine acyltransferases. The same size mRNA was also detected in rat heart, skeletal muscle and islets of Langerhans, in keeping with the notion that, in this species at least, CPT II is the product of a single gene. Two additional lines of evidence support this view. First, using the liver

FIGURE 2.

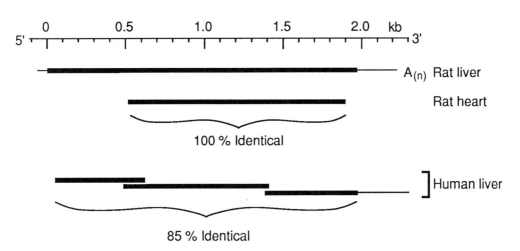

Comparison of rat and human CPT II cDNAs.
The solid bars represent coding regions. Percent identity is relative to the rat liver coding sequence.

FIGURE 3.

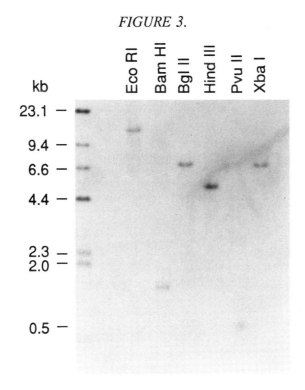

Genomic Southern blot of rat DNA probed with a labeled 3' fragment of CPT II cDNA. Rat DNA was digested with the indicated restriction endonucleases prior to analysis.

cDNA to screen a rat heart cDNA library (constructed in the vector pCDM8 by random hexamer priming and kindly provided by Dr. Kenneth R. Chien), a positive clone containing a 1.4kb insert was isolated. This was found to be 100% identical in nucleotide sequence to the corresponding region of the liver cDNA (Figure 2). Second, rat genomic DNA was treated with a variety of restriction endonucleases, followed by Southern analysis using a labeled 300 bp fragment from the 3' end of our CPT II cDNA. As seen from Figure 3, the simplicity of the hybridization pattern is indicative of a single copy gene for CPT II.

B. HUMAN CPT II

Screening of a human liver cDNA library (constructed in the vector λgt 11 by random hexamer priming and kindly provided by Drs. James Ou and William Rutter) with labeled rat CPT II cDNA yielded three positive clones. Sequence analysis established that together they represented an open reading frame corresponding to 98% of the coding sequence of the rat cDNA (Figure 2). The most 5' clone stopped short of the translation initiation site in the rat cDNA by 46 nucleotides. The most 3' clone contained a termination codon in the same position as that in the rat cDNA, followed by a 401bp untranslated region that terminated prior to the poly A$^+$ tail. The two nucleotide sequences are aligned in Figure 4 and show 85% identity over the coding region.

While this work was in progress Finnochiaro et al[23] reported the isolation of a full length cDNA for human liver CPT II that predicted a precursor protein of 658 amino acids, i.e., exactly the same number as in rat CPT II.[20] Moreover, because they were able to obtain an N-terminal sequence of the mature enzyme it was possible to define the cleavage site of the mitochondrial targeting peptide as occurring between leucine-25 and serine-26 of the nascent protein. As noted above, the leader sequence of rat CPT II is removed at precisely the same position. The amino acid sequences for human CPT II deduced by Finnochiaro et al[23] and by ourselves are compared in Figure 5. The broad agreement is self evident, discrepancies occurring in only 3 of 633 residues in the mature proteins. Despite these similarities, we found the mRNA encoding human CPT II to be substantially larger than its counterpart in the rat. This is illustrated in Figure 6, which shows the message size to be the same (~3kb) in human liver and fibroblasts, i.e. ~0.5kb larger than the rat mRNA. A similar size for CPT II mRNA in human liver was reported by Finnochiaro et al.[23] Whether the 5'- as well as the 3'-untranslated region of the human mRNA is longer than its rat equivalent remains to be established.

Curiously, although rat and human CPT II are both predicted to contain the same number of amino acids and to differ in molecular weight by only ~100Da (rat being the larger), their migration positions on 8% SDS gels suggested a much greater difference.[19] The possibility is raised that the two proteins undergo differential post-

```
   1   ATG ATG CCG CGC CTG CTG TTT CGT GCC TGG CCC CGC TGC CCC TCG CTT GTC CTG GGA GCC CCT AGT AGG CCC TTA AGT GCT GTC TCG GGG   RAT
  47                                    TT GGT CCG GGA GCC CCC AGT CGG CCC CTC AGC GCC GGC TCC GGG   HUMAN
                                        •• ••                    •  • •  • •  • •  • •

  91   CCG GAC GAT TAT CTG CAG CAC AGC ATC GTG CCC ACC ATG CAC TAC CAG GAC AGC CTG CCC AGG CTG CCT ATC CCT AAA CTT GAA GAC ACC   RAT
  91   CCC GGC CAG TAC CTG CAG CGC AGC ATC GTG CCC ACC ATG CAC TAC CAG GAC AGC CTG CCC AGG CTG CCT ATT CCC AAA CTT GAA GAC ACC   HUMAN
       •  • ••  •      •                                                              •   •  •

 181   ATG AAG AGA TAC CTC AAT GCA CAG AAG CCT CTC TTG GAT GAC AGC CAG TTC AGG AGA ACA GAA GCG TTG TGT AAG AAT TTT GAG ACT GGC   RAT
 181   ATT AGG AGA TAC CTC AGT GCA CAG AAG CCT CTC TTG AAT GAT GGC CAG TTC AGG AAA ACA GAA CAA TTT TGC AAG AGT TTT GAA AAT GGG   HUMAN
       •  • •                •        •                •      • •       •  •             •   •  •   •       •       • •   • •

 241   GTT GGG AAG GAG CTG CAT GCC CAC CTG CTT GCT CAG GAT AAG CAG AAT AAG CAC ACC AGC TAC ATC TCA GGC CCC TGG TTC GAT ATG TAT   RAT
 241   ATT GGA AAA GAA CTG CAT GAG CAG CTG GTT GCT CTG GAC AAA CAG AAT AAA CAT ACA AGC TAC ATT TCG GGA CCC TGG TTT GAT ATG TAC   HUMAN
       •  •  • •   •        •   ••  •      • •      • •   •   •       •    • •  •          •   •  •  • •          •           •

 361   TTA ACT GCT CGA GAC TCT ATC GTT TTA AAC TTT AAT CCA TTT ATG GCA TTC AAT CCG GAC CCA AAG TCT GAG TAT AAT GAC CAG CTT ACC   RAT
 361   CTA TCT GCT CGA GAC TCC GTT GTT TTA AAC TTT AAT CCA TTT ATG GCT TTC AAT CCT GAC CCA AAA TCT GAG TAT AAT GAC CAG CTC ACC   HUMAN
       • •  •            •   • •  •                            •         •        •            •                        •

 451   AGG GCA ACC AAC TTG ACT GTT TCT GCA GTC CGG TTT CTG AAG ACA CTT CAG GCT GGC CTT TTA GAA CCA GAA GTG TTC CAC TTG AAC CCT   RAT
 451   CGG GCA ACC AAC ATG ACT GTT TCT GCC ATC CGG TTT CTG AAG ACA CTC CGG GCT GGC CTT TTA GAG CCA GAA GTG TTC CAC TTG AAC CCT   HUMAN
       •                •              •  ••                        •  • •                    •

 541   TCC AAA AGT GAC ACA GAC GCC TTC AAA AGA CTC ATT CGC TTT GTT CCT CCC TCT CTG TCC TGG TAT GGA GCC TAC CTG TCC AAC GCA TAT   RAT
 541   GCA AAA AGT GAC ACT ATC ACC TTC AAG AGA CTC ATA CGC TTT GTG CCT TCC TCT CTG TCC TGG TAT GGG GCC TAC CTG TCC AAT GCG TAT   HUMAN
       ••            •  ••  •         •              •        •     •                        •                   •   •

 631   CCC CTG GAT ATG TCC CAG TAT TTC CGG CTT TTC AAT TCA ACT CGT ATA CCC AGA CCC AAT CGT GAT GAA CTC TTT ACT GAC ACC AAG GCT   RAT
 631   CCC CTG GAT ATG TCC CAG TAT TTT CGG CTT TTC AAC TCA ACT CGT TTA CCC AAA CCC AGT CGG GAT GAA CTC TTC ACT GAT GAC AAG GCC   HUMAN
                                •         •           •   •        •           •   •       •          • ••         •

 721   AGA CAC CTC CTG GTC CTA AGA AAA GGA CAT TTC TAT GTC TTT GAT GTC CTC GAT CAA GAT GGG AAC ATT GTG AAT CCC TTA GAG ATT CAG   RAT
 721   AGA CAC CTC CTG GTC CTA AGG AAA GGA AAT TTT TAT ATC TTT GAT GTC CTG GAT CAA GAT GGG AAC ATT GTG AGC CCC TCG GAA ATC CAG   HUMAN
                                •         • •   •   •              •                                    ••      • •  •    •

 811   GCA CAT CTA AAG TAC ATT CTC TCA GAC AGC AGC CCT GTG CCC GAG TTT CCT GTG GCA TAT CTG ACC AGT GAG AAC CGA GAT GTC TGG GCA   RAT
 811   GCA CAT CTG AAG TAC ATT CTC TCA GAC AGC AGC CCC CCC GAG TTT CCC CTG GCA TAC CTG ACC AGT GAG AAC CGA GAC GTC TGG GCA   HUMAN
                •                                  •  ••          •  ••                •                        •

 901   GAG CTC AGA CAG AAG TTG ATT TTT GAT GGC AAT GAG GAA ACC CTG AAG AAA GTG GAC TCT GCT GTC TTC TGC CTC TGC CTA GAT GAC TTC   RAT
 901   GAG CTC AGG CAG AAG CTG ATG AGT AGT GGC AAT GAG GAG AGC CTG AGG AAA GTG GAC TCG GCA GTG TTC TGT CTC TGC CTA GAT GAC TTC   HUMAN
                •         •   •  •  •            •   • •       •          •   •  •     •  •                  •

 991   CCG ATG AAG GAT CTT ATT CAC TTG TCA CAC ACC ATG CTG CAC GGC GAT GGC ACA AAC CGC TGG TTT GAT AAG TCC TTT AAC CTC ATT GTA   RAT
 991   CCC ATT AAG GAC CTT GTC CAC TTG TCC CAC AAT ATG CTG CAT GGG GAT GGC ACA AAC CGC TGG TTT GAT AAA TCC TTT AAC CTC ATT ATC   HUMAN
       •   ••       •        •           •        ••         •   •                                   •                     ••

1081   GCC GAG GAC GGC ACT GCT GCA GTC CAC TTC GAG CAC TCG TGG GGG GAT GGG GTG GCA GTG CTT AGG TTT TTC AAT GAA GTG TTC AGA GAC   RAT
1081   GCC AAG GAT GGC TCT ACT GCC ATC CAC TTT GAG CAC TCT TGG GGT GAT GGT GTG GCA GTG CTC AGA TTT TTT AAT GAA GTA TTT AAA GAC   HUMAN
           •   •       • •     •  •       •           •       •       •         •   • •      •       •               •   •  •

1171   AGC ACT CAG ACC CCT GCC ATC ACT CCC CAG AGC CAG CCT GCC GCC ACC AAC TCC TCT GCG TCT GTG GAG ACG CTC AGC TTC AAC CTC AGC   RAT
1171   AGC ACT CAG ACC CCT GCC GTC ACT CCA CAG AGC CAG CCA GCT ACC ACT GAC TCT ACT GCG ACG GTG CAG AAA CTC AAC TTC GAG CTG ACT   HUMAN
                               •       •           •      •   • •  • •   •• •     •   ••    •     • •       •        •  •   •

1261   GGT GCT CTC AAG GCT GGC ATC ACT GCT GCC AAG GAG AAG TTC GAC ACC ACG GTG AAA ACC CTC AGC ATT GAC TCC ATT CAG TTC CAG AGA   RAT
1261   GAT GCC TTA AAG ACT GGC ATC ACA GCT GCT AAG GAA AAG TTT GAT GCC ACC ATG AAA ACC CTC AGC ATT GAC TGC TCG CAG TTT CAG AGA   HUMAN
       •  • • •      • •          •        •      •      •   •  • ••          •                        • •  •       •

1351   GGT GGC AAG GAG TTC CTG AAG AAG AAG CAG CTG AGC CCC GAT GCG GTG GCC CAG CTG GCC TTC CAG ATG GCC TTC CTG AGA CAG TAT GGC   RAT
1351   GGA GGC AAA GAA TTC CTG AAG AAG CAA AAG CTG AGC CCT GAC GCA GTT GCC CAG CTG GCA TTC CAG ATG GCC TTC CTG CGG CAG TAC GGG   HUMAN
       •      •   • •                    •  •           •   •  •   ••           •              •              •       •   •

1441   CAG ACG GTG GCT ACC TAT GAG TCC TGC AGC ACT GCA GCA TTC AAG CAC GGC CGC ACA GAG ACT ATC CGC CCA TCC ATC TTT ACT AAG   RAT
1441   CAG ACA GTG GCC ACC TAC GAG TCC TGT AGC ACT GCC GCA TTC AAG CAC GGC CGC ACT GAG ACC ATC CGC CCG GCC TCC GTC TAT ACA AAG   HUMAN
           •        •       •        •           •      •               •       •   •          •   • •     • •  •    •

1531   AGA TGC TCC GAG GCG TTT GTC AGG GAC CCC TCC CAC CAC AGT GTG GGC GAG CTT CAG CAC ATG ATG GCT GAG TGT TCC AAA TAC CAT GGC   RAT
1531   AGG TGC TCT GAG GCC TTT GTC AGG GAG CCC TCC AGG CAT AGT GTG GCT GGT GAG CTT CAG CAG ATG ATG GTT GAG TGC TCC AAG TAC CAT GGC   HUMAN
       •       •       •               •      •   ••          •       •              •              •      •       •

1621   CAG CTG ACC AAA GAA GCA GCG ATG GGC CAG GGC TTT GAC CGA CAC TTG TAT GCT CTG CGC TAT CTG GCA ACG GCC AGA GGA CTC AAC CTA   RAT
1621   CAG CTG ACC AAA GAA GCA ATG GGC CAG GGC TTT GAC CGA CAC TTG TTT GCT CTG CGC CAT CTG GCA GCC AAA GGG ATC ATC TTG   HUMAN
                               •                                   •              •          • •      • •  •    • •  • •

1711   CCT GAG CTC TAT CTG GAT CCT GCA TAC CAG CAG ATG AAC CAC AAC ATC CTG TCC ACC AGC ACT CTG AAC AGC CCA GCA GTG AGC CTT GGG   RAT
1711   CCT GAG CTC TAC CTG GAC CCT GCA TAC GGG CAG ATA AAC CAC AAT GTC CTG TCC ACG AGC ACA CTG AGC AGC CCA GCA GTG AAC CTT GGG   HUMAN
                       •        •           ••          •       •        •          •           •   •              •

1801   GGC TTT GCC CCT GTG GTC CCT GAT GGC TTT GGC ATT GCA TAT GCT GTT CAC GAT GAC TGG ATA GGC TGC AAT GTC TCC TCC TAC TCA GGA   RAT
1801   GGC TTT GCC CCT GTG GTC TCT GAT GGC TTT GGT GTT GGG TAT GCT GTT CAT GAC AAC TGG ATA GGC TGC AAT GTC TCT TCC TAC CCA GGC   HUMAN
                               •             •  ••  •              •      •             •          •        •

1891   CGC AAT GCC CGA GAG TTT CTC CAC TGT GTC CAG AAG TGC TTG GAA GAC ATT TTC GAT GCT CTA GAA GGC AAA GCC ATC AAA ACT TAG   RAT
1891   CGC AAT GCC CGG GAG TTT CTC CAA TGT GTG GAG AAG GCC TTA GAA GAC ATG TTT GAT GCC TTA GAA GGC AAA TCC ATC AAA AGT TAA   HUMAN
                •            •       ••         •  •       • •            •  • •      •  • •           •          • •  • •
```

Nucleotide sequence of rat and human CPT II cDNAs.

The three partial cDNAs for human CPT II (Figure 2) yielded an open reading frame of nucleotide sequence having 85% identity over the coding region of the rat cDNA. Non-identical bases are indicated by the dots.

FIGURE 5.

```
  1   MVPRLLLRAW  PRGPAVGPGA  PSRPLSAGSG  PGQYLQRSIV  PTMHYQDSLP

 51   RLPIPKLEDT  IRRYLSAQKP  LLNDGQFRKT  EQFCKSFENG  IGKELHEQLV

101   ALDKQNKHTS  YISGPWFDMY  LSARDSVVLN  FNPFMAFNPD  PKSEYNDQLT

151   RATNMTVSAI  RFLKTLRAGL  LEPEVFHLNP  AKSDTITFKR  LIRFVPSSLS

201   WYGAYLVNAY  PLDMSQYFRL  FNSTRLPKPS  RDELFTDDKA  RHLLVLRKGN

251   FYIFDVLDQD  GNIVSPSEIQ  AHLKYILSDS  SPGPEFPLAY  LTSENRDIWA
                                                  —A—

301   ELRQKLMSSG  NEESLRKVDS  AVFCLCLDDF  PIKDLVHLSH  NMLHGDGTNR

351   WFDKSFNLII  AKDGSTAVHF  EHSWEDGVAV  LRFFNEVFKD  STQTPAVTPQ
                        —I—        —G—

401   SQPATTDSTV  TVQKLNFELT  DALKTGITAA  KEKFDATMKT  LTIDCVQFQR

451   GGKEFLKKQK  LSPDAVAQLA  FQMAFLRQYG  QTVATYESCS  TAAFKHGRTE

501   TIRPASVYTK  RCSEAFVREP  SRHSAGELQQ  MMVECSKYHG  QLTKEAAMGQ

551   GFDRHLFALR  HLAAAKGIIL  PELYLDPAYG  QINHNVLSTS  TLSSPAVNLG

601   GFAPVVSDGF  GVGYAVHDNW  IGCNVSSYPG  RNAREFLQCV  EKALEDMFDA

651   LEGKSIKS*
```

The deduced amino acid sequence of human CPT II.
The amino acid sequence shown is that reported by Finnochiaro et al.[23] The underlined sequence is that deduced from our cDNA shown in Figure 4. The two differ in 3 residues as indicated. The arrow denotes the cleavage site of the mitochondria leader peptide.

FIGURE 6.

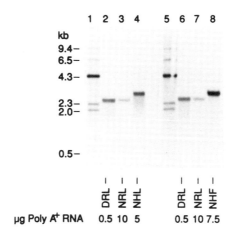

Northern analysis of poly A^+ RNA from rat and human tissues.
The indicated amounts of poly A^+ RNA were analyzed using a labeled (partial) human cDNA as probe. NRL, normal rat liver; DRL, liver from DEHP-treated rats; NHL, normal human liver; NHF, normal human fibroblasts. Lanes 1 and 5 show DNA size standards.

translational modification. Alternatively, the 20% variation in their amino acid make up results in anomalous behavior of one or the other on SDS-PAGE. Interestingly, the enzyme from mouse and monkey liver comigrated with the human, not the rat isoform.[19]

Because of its much greater hydrophobicity and detergent lability, CPT I has proved to be extremely difficult to isolate in pure form. Efforts in this direction, with a view to cloning and sequencing a corresponding cDNA, are currently under way.

IV. TOPOGRAPHICAL FEATURES OF THE CPT ISOZYMES

Even if cDNA-deduced primary amino acid sequences provide insight into higher order structures of CPT I and CPT II, such information alone might be of limited value in understanding the topographical relationships between the two proteins and their respective membrane environments. Nor would it necessarily elucidate the peculiar properties of CPT I, such as its unique inhibitability by malonyl-CoA or TG-CoA[3] and its changing sensitivity to the former agent observed both in vivo and in vitro under certain conditions.[25-28] To explore these aspects of the system we recently conducted a series of experiments involving exposure of mitochondria to a variety of proteases, followed by measurement of CPT I and CPT II activity in the absence and presence of malonyl-CoA or etomoxir-CoA (equivalent to TG-CoA in this context). Marker enzymes chosen were adenylate kinase (inter-membrane space), β-hydroxybutyrate dehydrogenase (inner membrane) and glutamate dehydrogenase (matrix). We summarize here the main findings, beginning with the effects of three of the proteases tested on CPT I of rat liver and heart mitochondria.

A. EFFECTS OF PROTEASES ON CPT I OF LIVER AND HEART MITOCHONDRIA

It was known, principally from the work of Zammit et al,[29] that in mitochondria prepared at low temperature from fed rats, CPT I exhibits high sensitivity to malonyl-CoA, but becomes desensitized to the inhibitor with a rise in temperature. Moreover, this spontaneous desensitization could be blocked, or reversed, by inclusion of malonyl-CoA in the incubations. A similar phenomenon occurs in vivo with alteration in nutritional status, and also with the induction and reversal of diabetes.[25-28] These changes in malonyl-CoA sensitivity of CPT I appear to result, at least in part, from alterations in fluidity of the mitochondrial outer membrane.[30]

We observed that exposure of liver mitochondria (prepared at 0-4° from fed rats) to trypsin, chymotrypsin or nagarse at room temperature caused a time-dependent destruction of CPT I activity. In addition, particularly in the case of chymotrypsin and nagarse, there was a marked acceleration in malonyl-CoA desensitization of the enzyme during the early stages of protease action. There was no effect, however, on the ability of etomoxir-CoA to cause complete (and irreversible) inhibition of enzyme activity.

Inclusion of malonyl-CoA during the incubation not only prevented spontaneous desensitization of CPT I (as expected), but also greatly attenuated both the protease-induced desensitization and destruction of the enzyme. When added after protease action had begun, malonyl-CoA prevented further loss of CPT I activity, but did not cause resensitization of the desensitized enzyme to the inhibitor.

Unlike liver CPT I, the enzyme in heart mitochondria showed very little spontaneous desensitization to malonyl-CoA during the $0° \rightarrow 22°$ transition. The effects of proteases, however, were similar to those seen with the liver system, i.e., destruction of CPT I activity coupled with a marked loss of malonyl-CoA sensitivity, trypsin again being the least effective in terms of the latter response. Also, regardless of treatment, etomoxir-CoA caused complete enzyme inhibition.

Even in incubations carried out at $0°$, nagarse had the effect of decreasing both CPT I activity and its malonyl-CoA sensitivity in liver mitochondria. Under these conditions heart CPT I responded differently; it retained full catalytic activity but still became desensitized to malonyl-CoA. Similar findings were reported by Murthy and Pande[31] who interpreted them to mean that CPT I spans the mitochondrial outer membrane, with its malonyl-CoA binding site (nagarse sensitive) oriented externally and the palmitoyl-CoA/TG-CoA binding domain (nagarse insensitive) facing the inter-membrane space. Such a formulation is difficult to reconcile with the fact that malonyl-CoA and TG-CoA/etomoxir-CoA appear to interact at a common site, and that in all cases inhibitor binding is antagonized by palmitoyl-CoA.[3]

An alternative model that encompasses our previous views on CPT I-inhibitor interactions,[3] the effects of proteases on the system, and recent suggestions by Kolodziej and Zammit[30] concerning shifts in malonyl-CoA sensitivity of the enzyme, is presented in Figure 7. In this scheme, CPT I is seen to exist in a phospholipid milieu (critical for activity) as an equilibrium mixture of two conformations, A and B. Because the relationship between sites 1 and 3 (the palmitoyl-CoA and carnitine binding sites, respectively) is the same in A and B, both forms are equally active. In the presence of malonyl-CoA, form B is converted into form A', which now contains the inhibitor bound to site 1 through its CoA moiety and to site 2a via the carboxyl function on carbon 2. (We suspect that site 2a might be a histidine residue, which would be more positively charged as the pH is lowered over the neutral range and thus exert greater attraction on the carboxyl group of malonyl-CoA. This would account for the tighter binding and greater potency of the inhibitor at pH 6.8 compared with 7.6[3]). If the mitochondria are now resuspended in fresh medium at $0°$, the inhibitor dissociates from the enzyme, which is now "frozen" in form A. The greater malonyl-CoA sensitivity of A compared with B relates to the closer apposition of sites 1 and 2a in the former conformation. With a subsequent rise in temperature, membrane fluidity increases and, concomitantly, form A returns to form B. In the absence of malonyl-CoA, both A and B are susceptible to attack by proteases at all sites, resulting in loss of catalytic activity as well as malonyl-CoA binding. In the case of form A, should protease action at site 2a precede that at site

1 (or 3), malonyl-CoA sensitivity would be lost more rapidly than is enzyme function. When malonyl-CoA is bound (form A') all of the functional domains are protected from protease action.

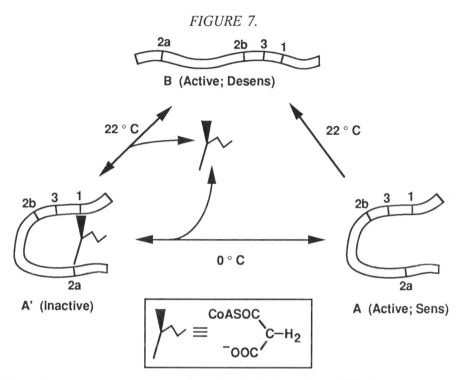

FIGURE 7.

Model for effects of temperature and malonyl-CoA on liver CPT I.
1, binding site for thioester group of palmitoyl-CoA, malonyl-CoA, or TG-CoA; 2a, binding site for carboxyl group of malonyl-CoA; 2b, binding site for epoxide ring of TG-CoA; 3, binding site for carnitine. Sens, sensitized; Desens, desensitized.

In contrast to the situation with malonyl-CoA, agents such as TG-CoA or etoxomir-CoA are viewed as binding to site 1 (again through their thioester group) and covalently to site 2b by opening of the epoxide ring on carbon 2. If in the A⇌B interconversion, or during protease attack, the relationship between sites 1 and 2b does not change, enzyme activity would be completely inhibited by such agents under all of the conditions discussed.

In heart mitochondria the transition of CPT I from form A to form B does not occur either in response to temperature in vitro or , apparently, during starvation in vivo.[32] Presumably, this reflects differences in outer membrane composition between liver and heart mitochondria and/or the fact that CPT I exists as distinct isozymes in the two tissues.[19]

We emphasize that the scheme depicted in Figure 7 is but one of several that might be considered to accommodate the known features of CPT I. At present it represents a minimal construct since it does not take into account possible protein-protein interactions that might be important in the function and/or regulation of the enzyme. Nor does it allow speculation on the orientation of the transferase within the membrane. Nevertheless, it is amenable to further experimental testing and could provide a useful working model for future attempts to unravel the complexities of this intriguing system. In this regard, the successful isolation and sequencing of a cDNA for CPT I will be of immense importance.

B. EFFECTS OF PROTEASES ON CPT II.

In the studies summarized above, trypsin, chymotrypsin and nagarse destroyed not only CPT I activity but also that of adenylate kinase and β-hydroxybutyrate dehydrogenase. They were totally without effect, however, on the activity of CPT II or glutamate dehydrogenase. This was true both in liver and heart mitochondria. The implication is that while the action of all three proteases extended to a point in the inner membrane housing β-hydroxybutyrate dehydrogenase, they did not reach the catalytic center of CPT II, which presumably faces the matrix compartment. In keeping with this interpretation was the finding that all three agents readily destroyed both CPT I and CPT II in freeze-thawed mitochondria.

V. EPILOGUE

In its original formulation the CPT system was proposed to consist of two membrane-bound enzymes catalyzing the same reaction in opposite direction on either side of a substrate barrier. Thirty years later the same concept, possibly unique in biology, still appears to be valid. Understanding of how this complex transport mechanism actually works, though slow in beginning and still far from complete, is now evolving at an exciting pace. Not surprisingly, much confusion and healthy debate has been generated in the process. However, now that the powerful tools of molecular biology are beginning to be applied in this area it is predictable that major enlightenment will be rapidly forthcoming.

REFERENCES

1. Fritz, I. B. and Yue K. T. N., Long-chain carnitine acyltransferase and the role of acylcarnitine derivatives in the catalytic increase of fatty acid oxidation induced by carnitine, J. Lipid Res. 4,279-288, 1963.
2. Bremer, J., Carnitine in intermediary metabolism. The biosynthesis of palmitoylcarnitine by cell subfractions, J. Biol. Chem. 238, 2774-2779, 1963.
3. McGarry, J. D., Woeltje, K. F., Kuwajima, M. and Foster, D. W., Regulation of ketogenesis and the renaissance of carnitine palmitoyltransferase, Diabetes Metab. Rev., 5, 271-284,1989.

4.   Tutwiler, G. F., Brentzel, H.J. and Kiorpes, T. C., Inhibition of mitochondrial carnitine palmitoyltransferase A in vivo with methyl 2-tetradecylglycidate (methyl palmoxirate) and its relationship to ketonemia and glycemia, Proc. Soc. Exp. Biol. Med. 178, 288-296, 1985

5.   Wolf, H. P. O. and Engel, D. W., Decrease of fatty acid oxidation, ketogenesis and gluconeogenesis in isolated perfused rat liver by phenylalkyl oxirane carboxylate (B 807-27) due to inhibition of CPT I (EC 2.3.1.21) Eur J. Biochem., 146, 359-363, 1985

6.   Tein, I., Demaugre, F., Bonnefont, J. P. and Saudubray, J.M., Normal muscle CPT I and CPT II activities in hepatic-presentation patients with CPT I deficiency in fibroblasts. Tissue specific isoforms of CPT I?, J. Neurol. Sci., 92, 229-245, 1989

7.   Demaugre, F., Bonnefont, J. P., Colonna, M., Cepanec, C., Leroux, J. P. and Saudubray, J. M., Infantile form of carnitine palmitoyltransferase II deficiency with hepatomuscular symptoms and sudden death, J. Clin. Invest., 87, 859-864, 1991

8.   Pande, S. V., A mitochondrial carnitine acylcarnitine translocase system, Proc Natl. Acad. Sci. USA, 72, 883-887, 1975

9.   Murthy, M. S. R., Pande, S. V., Malonyl-CoA binding site and the overt carnitine palmitoyltransferase activity reside on the opposite sides of the outer mitochondrial membrane, Proc. Natl. Acad. Sci. USA, 84, 378-382, 1987

10.   Lund, H. and Woldegiorgis, G., Carnitine palmitoyltransferase: separation of enzyme activity and malonyl-CoA binding in rat liver mitochondria, Biochim. Biophys. Acta, 878, 243-249, 1986

11.   Bieber, L. L., Carnitine, Annu. Rev. Biochem., 57, 261-283, 1988

12.   Zammit, V. A., Corstorphine, C. G. and Kolodziej, M. P., Target size analysis by radiation inactivation of carnitine palmitoyltransferase activity and malonyl-CoA binding in outer membranes from rat liver mitochondria, Biochem. J., 263, 89-95, 1989

13.   Kerner, J. and Bieber, L., Isolation of a malonyl-CoA-sensitive CPT/β-Oxidation enzyme complex from heart mitochondria, Biochemistry, 29, 4326-4334, 1990

14.   Murthy, S. R. and Pande, S. V., Characterization of a solubilized malonyl-CoA sensitive carnitine palmitoyltransferase from the mitochondrial outer membrane as a protein distinct from the malonyl-CoA insensitive carnitine palmitoyltransferase of the inner membrane, Biochem. J., 268, 599-604, 1990

15.   Ghadiminejad, I. and Saggerson, E. D., The relationship of rat liver overt carnitine palmitoyltransferase to the mitochondrial malonyl-CoA binding entity and to the latent palmitoyltransferase, Biochem. J., 270, 787-794, 1990

16.   Declercq, P.E., Venincasa, M.D., Mills, S. E., Foster, D. W. and McGarry, J. D., Interactions of malonyl-CoA with mitochondrial carnitine palmitoyltransferase. J. Biol. Chem., 260, 12516-12522, 1985

17. Declercq, P. E., Falck, J. R., Kuwajima, M., Tyminski, H.,Foster, D. W. and McGarry, J. D., Characterization of the mitochondrial carnitine palmitoyltransferase enzyme system. I. Use of inhibitors. J. Biol. Chem., 262, 9812-9821, 1987

18   Woeltje, K. F., Kuwajima, M., Foster, D. W. and McGarry, J.D., Characterization of the mitochondrial carnitine palmitoyltransferase enzyme system. II. Use of detergents and antibodies, J. Biol. Chem., 262, 9822-9827, 1987

19. Woeltje, K. F., Esser, V., Weis, B. C., Cox, W. F., Schroeder, J. G., Liao, S. T., Foster, D. W. and McGarry, J. D., Inter-tissue and inter-species characteristics of the mitochondrial carnitine palmitoyltransferase enzyme system, J. Biol. Chem., 18, 10714-10719, 1990

20. Woeltje, K. F., Esser, V., Weis, B.C., Sen, A., Cox, W., McPhaul, M. J., Slaughter, C. A., Foster, D. W. and McGarry, J. D., Cloning, sequencing and expression of a cDNA encoding rat liver mitochondrial carnitine palmitoyltransferase II, J. Biol. Chem., 18, 10720-10725, 1990

21. Brady, P.S., Feng, Y. and Brady, L. J., Transcriptional regulation of carnitine palmitoyltransferase synthesis in riboflavin deficiency, J. Nutr., 118, 1128-1136, 1988

22. Brady, P. S., Liu, J. S., Park, E. A., Hanson, R. W. and Brady, L. J., Isolation and characterization of the promoter for the gene coding for carnitine palmitoyltransferase, FASEB J., 5, A817, 1991

23. Finocchiaro, G., Taroni, F., Rocchi, M., Martin, A. L., Colombo, I., Tarelli, G. T., DiDonato, S., cDNA cloning, sequence analysis, and chromosomal localization of the gene for human carnitine palmitoyltransferase, Proc. Natl. Acad. Sci. USA, 88, 661-665, 1991

24. McGarry, J. D., Sen, A., Esser, V., Woeltje, K. F., Weis, B. C. and Foster, D. W., New insights into the mitochondrial carnitine palmitoyltransferase enzyme system, Biochimie, 1991, in press

25. Grantham, B. D. and Zammit, V. A., Restoration of the properties of carnitine palmitoyltransferase I in liver mitochondria during re-feeding of starved rats, Biochem. J., 239, 485-488, 1986

26. Cook, G. A. and Cox, K. A., Hysteretic behavior of carnitine palmitoyltransferase: The effect of preincubation with malonyl-CoA, Biochem. J., 236, 917-919, 1986

27. Grantham, B. D. and Zammit, V. A., Role of carnitine palmitoyltransferase I in the regulation of hepatic ketogenesis during the onset and reversal of chronic diabetes. Biochem. J., 249, 409-414, 1988

28. Prip-Buus, C., Pegorier, J.P., Duee, P. H., Kohl, C. and Girard, J., Evidence that the sensitivity of carnitine palmitoyltransferase I to inhibition by malonyl-CoA is an important site of regulation of hepatic fatty acid oxidation in the fetal and newborn rabbit, Biochem. J., 269, 409-415, 1990

29. Zammit, V. A., Corstorphine, C. G. and Gray, S. R., Changes in the ability of malonyl-CoA to inhibit carnitine palmitoyltransferase I activity and to bind to rat liver mitochondria during incubation in vitro, Biochem. J., 222, 335-342, 1984

30. Kolodziej, M. P. and Zammit, V. A., Sensitivity of inhibition of rat liver mitochondrial outer-membrane carnitine palmitoyltransferase by malonyl-CoA to chemical- and temperature-induced changes in membrane fluidity, Biochem. J., 272, 421-425, 1990

31. Cook, G.A and Lappi, M. D., Re-evaluation of malonyl-CoA inhibition carnitine palmitoyltransferase in myocardial mitochondria, FASEB J., 5, A711, 1991

DISCUSSION OF THE PAPER

G. WOLDEGIORGIS *(Madison, WI)*: How do you explain that adding a malonyl-CoA binding protein fraction (inactive CPT I) restores malonyl-CoA sensitivity to CPT II?

MCGARRY: As I have not had the opportunity to examine this data in detail it is difficult to offer a ready explanation. I would, however, wonder if it is possible that in such mixing experiments the inactive CPT I became partially reactivated such that the malonyl-CoA sensitivity observed relates to the resurrected CPT I rather than to CPT II.

G. COOK *(Memphis, TN)*: We have recently obtained data with proteinase treatment of mitochondria that are very similar to yours, but we have also worked with inhibitors of CPT other than malonyl-CoA. Nagarse treatment of CPT does not affect inhibition by 2-Bromo palmitoyl-CoA, gluburide, or adriamycin analogues as it affects malonyl-CoA inhibition. We believe this is because of binding of these inhibitors at different sites on CPT. Could you comment on this possibility?

MCGARRY: Our observation has been that while Nagarse greatly reduces malonyl-CoA sensitivity of CPT I, it does not alter the ability of etomoxir-CoA to inhibit the enzyme irreversibly. We have suggested that the explanation might lie in a differential effect of Nagarse on the sites required for the binding of the carboxyl group of malonyl-CoA and the epoxide ring of etomoxir-CoA. It seems possible that the inhibitors to which you refer might also interact with the enzyme at a Nagarse insensitive site (or at the catalytic center if this was also destroyed by the protease in your experiments).

M. MURTHY *(East Lansing, MI)*: Have you done the topographical studies using proteases on [^{3}H]-TG-CoA labeled CPT I in mitochondria?

MCGARRY: We have not carried out detailed studies along these lines. However, we have established that the [^{3}H]-TG-CoA binding protein (as seen on SDS gels) is lost after exposure of [^{3}H]-TG-CoA labeled mitochondria to agents such as Nagarse or chymotrypsin.

CARNITINE PALMITOYLTRANSFERASE

Charles L. Hoppel
Departments of Pharmacology and Medicine
Case Western Reserve University
School of Medicine
Medical Research Service
VA Medical Center
Cleveland, Ohio 44106

I. INTRODUCTION

Carnitine palmitoyltransferase (CPT) is an interesting but difficult enzyme to study.[1,2,3,4] Not only are we uncertain about how many CPTs exist, and where the enzymes are localized within the cell as well as within the mitochondria, but there is also no agreement on how optimally to measure the enzyme.

II. MEASUREMENT OF ENZYMATIC ACTIVITY

A. MODIFICATIONS OF ENZYMATIC ANALYSIS

A number of factors are known to be important in obtaining the maximal activity of CPT in a biological sample.[1,2] While liver mitochondria show an overt and latent activity, Zierz and Engel have reported that skeletal muscle mitochondrial CPT does not show these properties.[5] In attempting to optimize the incubation conditions for the measurement of CPT for mitochondrial studies and for diagnostic purposes in tissue homogenates, we have examined each of the components in the assay. The radioenzymatic assay to measure the forward reaction of CPT involves the determination of the conversion of radioactive carnitine into palmitoylcarnitine.[2] The optimal assay conditions we have found to measure total activity in rat and human skeletal muscle homogenates and mitochondria include: 80 mM KCl, 50 mM MOPS buffer, pH 7.0, 1 mM N-ethylmaleimide (NEM), 1 mM EGTA, 0.2 mM palmitoyl-CoA, and 15 mM L-[14C-Me]-carnitine. The high concentration of L-carnitine is required for maximal total activity in skeletal muscle, at least for rat and human.[6] The NEM is added to react with the product CoASH so that the reaction cannot be reversed. The addition of NEM leads to a small but reproducible increase in activity. The KCl is essential to maximize the activity of the outer CPT.[7] Under these conditions we observe a specific activity for CPT of about 50 mU/mg rat skeletal muscle mitochondrial protein. When the concentration of palmitoyl-CoA is 0.05 mM and freshly isolated skeletal muscle mitochondria are examined, the activity is about one-half of this maximal rate. We use Lubrol-PX as the detergent in the purification of the tightly membrane-bound inner CPT activity (CPT-B) in mitochondria. As a part of the examination of the effect of a group of detergents on the activity of mitochondrial CPT in liver mitochondria, we observed an approximate doubling of the maximum rate when Lubrol-PX was used. We therefore examined the effect of increasing concentrations of Lubrol-PX to the incubation medium in skeletal muscle mitochondria. As shown in Figure1 there is a 1.8-fold increase in the specific activity of CPT at 5 mM Lubrol-PX. Similar results are observed with human skeletal muscle mitochondria, as well as with skeletal muscle homogenates.

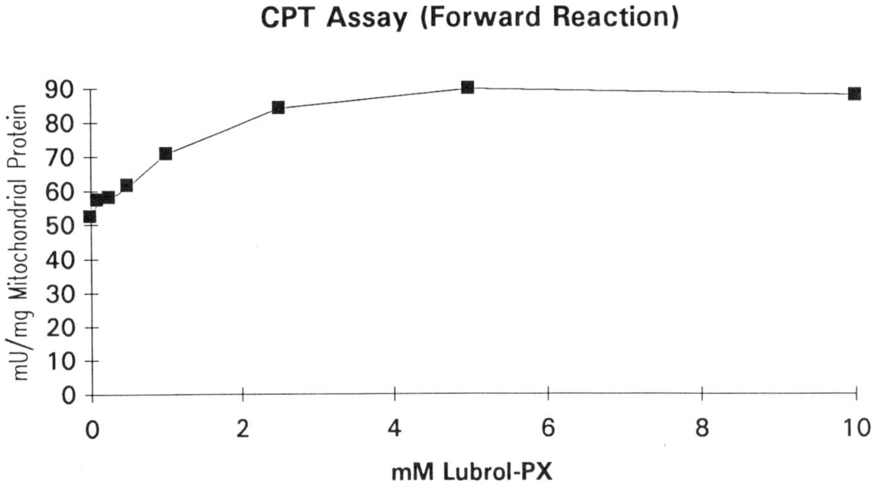

CPT Assay (Forward Reaction)

FIGURE 1.   The effect of increasing concentrations of Lubrol-PX on the specific activity of rat skeletal muscle mitochondria.

For maximal expression of CPT activity measured in the forward direction in tissue homogenates, isolated mitochondria, and cultured skin fibroblasts, Lubrol-PX is included in the assay.

B.   APPLICATION OF ASSAYS TO THE STUDY OF SKIN FIBROBLAST HOMOGENATES

The use of cultured skin fibroblasts in the evaluations of patients with disorders of fatty acid oxidation has become a standard procedure.   The activity of the forward reaction for CPT using homogenates of cultured skin fibroblasts is maximal under the conditions described above for isolated mitochondria. It has also been our custom to use the reverse reaction of CPT activity for diagnostic studies because the activity is 4-8 times faster than the forward reaction.[8]   This increases the sensitivity of the assay in detecting decreased amounts of activity or when only small amounts of tissue are available.

Malonyl-CoA sensitivity in the forward CPT reaction under modified assay conditions has been shown to be useful in the evaluation of patient tissues or cells.[9]   We have found that the order of addition of the components of the assay mixture is critical to obtain maximal malonyl-CoA sensitivity. The incubation mixture for the forward reaction given above is modified for these studies. It does not contain detergents, the concentration of palmitoyl-CoA is decreased to 0.05 mM, 4 mg bovine serum albumin (defatted) is added per mL, and the L-carnitine concentration is decreased to 0.5 mM. If the reaction is started with isolated mitochondria, we find little if any inhibition with 0.2 mM malonyl-CoA. However, if the isolated, intact mitochondria are preincubated with palmitoyl-CoA and malonyl-CoA for 2 minutes and the reaction started by the addition of L-carnitine, there is 70-80% inhibition of the overt CPT activity.

We have applied the above three types of assays for CPT activity to homogenates of cultured skin fibroblasts from controls and patients suspected of having CPT deficiency based on their clinical presentation. The homogenates of freshly harvested fibroblasts or frozen fibroblasts are sonicated for 3 seconds as a pretreatment to obtain optimal activity for the malonyl-CoA sensitivity experiments. Such sonication treatment is not necessary for the measurement of the forward or backward reactions.

TABLE 1.

Carnitine palmitoyltransferase activity in skin fibroblasts. The patients were suspected of having CPT deficiency based on clinical presentation. The control data are given as mean $\pm$ SEM.

Assay	CPT-A deficiency	CPT-B deficiency	Controls N = 14
Forward	2.36	0.049	2.760 $\pm$ 0.22
Backward	7.15	1.13	7.990 $\pm$ 0.56
Modified			
- Malonyl-CoA	0.195	0.509	0.911 $\pm$ 0.043
+ Malonyl-CoA	0.028	0.0	0.382 $\pm$ 0.022
delta	0.0	0.509	0.528 $\pm$ 0.032

The data shown in Table 1 were obtained from skin fibroblasts from a patient diagnosed as having CPT-B [2 or inner] deficiency,[10] a patient with CPT-A [1 or outer] deficiency (kindly provided by Dr. Steve Cederbaum) and from 14 controls (mean and SEM given). The patient described as having CPT-B deficiency shows about 20% of the normal total CPT activity measured in the forward and the backward direction. All of the CPT activity measured in the modified assay with malonyl-CoA present is sensitive. It would appear that the residual activity measured in the forward reaction represents only malonyl-CoA sensitive activity. In contrast to this reasonably straight-forward use of the assays, when the forward and backward reaction are measured using skin fibroblasts from the patient with CPT-A deficiency, the total activity measured is 85-90% of control values. These total activities are within the range that would be considered normal. However, when the sensitivity to malonyl-CoA is examined, the activity measured in this type of assay shows no sensitivity to malonyl-CoA. In skin fibroblasts it appears that CPT-A (malonyl-CoA sensitive CPT) contributes only 10-20% of the total activity when measured in either the forward or backward direction.

The measurement of total activity provides a clear definition for possible decreases in CPT-B activity, whereas total activity cannot be used to define changes in CPT-A activity. For determining deficiencies in CPT-A it is essential to expand the approach at least to include malonyl-CoA sensitivity. When CPT-A has been characterized and probes are available to detect the protein, it will be possible to establish this purported distribution of the enzymatic activities and refine the ability to detect CPT-A deficiencies.

III. SUBMITOCHONDRIAL LOCALIZATION OF CPT ACTIVITY

A. CURRENT STATUS

The demonstration of overt and latent CPT activity in liver mitochondria led to the proposal of two CPT activities (one outer and another inner) to transport acyl groups across the mitochondrial inner membrane. The studies using digitonin to fractionate the mitochondria showed that CPT activity was present on the outer surface of the inner membrane of rat liver mitochondria, and another activity was present on the inner surface of that membrane.[8] Phospholipase C was used to enhance the fractionation of beef liver mitochondria, and a similar localization for two CPT activities was reported.[11] Phosphate-induced swelling also has been used to fractionate the membranes of rat liver mitochondria.[12,13,14] Using phosphate swelling, Murthy and Pande demonstrated that the malonyl-CoA sensitive CPT activity in rat liver mitochondria was enriched in the outer membrane fraction.[15] In a series of studies, they reported the malonyl-CoA sensitive CPT activity present on the inner surface of the outer membrane, whereas the malonyl-CoA sensitive conferring properties were exposed on the outer surface.[15,16,17] Using the same phosphate swelling procedure as Murthy and Pande, we have reproduced their data on the distribution of enzymatic activity. However, the enrichment of malonyl-CoA sensitive CPT activity in the outer membrane was only 20% of the enrichment of monoamine oxidase activity, (the marker for the outer membrane). If the malonyl-CoA sensitive CPT activity were localized exclusively on the outer membrane, concordance of the enrichment should be expected. This apparent lack of agreement between the two enzyme activities led us to investigate other methods of mitochondrial membrane fractionation.

B. THE USE OF THE FRENCH PRESS TO STUDY CPT LOCALIZATION

Our intent was to find another procedure that used a fractionation technique that did not use chemicals (digitonin), enzymes (Phospholipase C), or produce the disruption that occurs with phosphate-induced swelling. Decker and Greenawalt reported the use of the French Press to fractionate mitochondria.[18,19] Their procedure is dependent on the use of a hyperosmolar medium and the shear force produced by the hydraulic press, which ruptures the outer membrane. They were able to isolate an inner membrane-matrix preparation (IM-M) that retained both good respiratory control ratio and coupled oxidative phosphorylation in addition to an outer membrane fraction. This procedure offers many advantages over other methods, yet it is interesting that it has been rarely used in mitochondrial studies. However, we selected this procedure to attempt to answer the questions raised concerning the location of the malonyl-CoA sensitive CPT activity.

An additional concern for studies involving CPT and in particular the malonyl-CoA sensitive activity is the recent observations that both the microsomes[20] and peroxisomes[21] contain carnitine acyltransferases activities that are sensitive to malonyl-CoA. Therefore, we subjected isolated intact rat liver mitochondria to a self-forming Percoll gradient to prepare a mitochondrial fraction with minimal contamination by peroxisomes. These freshly prepared Percoll-purified rat liver mitochondria were then processed according to the procedure for membrane fractionation using the French Press.[18,19] We varied the pressure generated by the French Press and found that in our experimental system a pressure of 2000 psi disrupted the outer membrane so that 85% of the outer membrane was released from the low speed inner membrane-matrix (IM-M) preparation (Table 2). When the low speed supernatant was subjected to ultracentrifugation, the membrane fraction contained 55% and the soluble supernatant fraction contained 30% of the mitochondrial outer membrane marker monoamine oxidase. The first order rate constant was determined for cytochrome C oxidase as a marker of the mitochondrial inner membrane (cristae membranes) in all fractions, and this was used to calculate recoveries. Only 6% of the cytochrome C oxidase activity was released and this was recovered in the high speed membrane fraction. Citrate synthase (a mitochondrial matrix marker) was used to determine the degree of inner membrane disruption and only 7.7% of the activity was released. CPT activity was measured in both the forward and backward reaction. Of the activity in the Percoll-purified mitochondria, only 1.2-1.4% of the total activity was released and recovered (predominantly in the soluble fraction). Only 0.2% of the beginning CPT activity was present in the outer membrane fraction which contained 55% of the monoamine oxidase activity. These data do not support the notion that CPT activity is present in the outer membrane.

TABLE 2.

Fractionation of rat liver mitochondria using the French Press. The procedure described by Decker and Greenawalt[18,19] was used. Data are expressed as the percentage recovered in a particular fraction with the beginning Percoll-purified mitochondria as 100%.

	Inner membrane matrix	Supernatant membrane	Outer	Total
Protein	58.6	15.7	5.9	80.2
Monoamine oxidase	16.6	30.0	55.6	103.2
Citrate Synthetase	93.3	4.4	3.3	101.0
Cytochrome C oxidase	92.8	0.0	6.2	99.0
CPT-forward	99.2	1.0	0.2	100.4
CPT-backward	96.0	1.2	0.2	97.4

The data in Table 2 are presented as percent of activity present in the Percoll-purified mitochondria used for the French Press experiment. We also measured malonyl-CoA sensitive CPT activity in all of the fractions. Table 3 contains data showing the specific activity of the total CPT activity measured by the forward reaction and of the malonyl-CoA sensitive activity. Using this type of presentation of the specific activity of CPT, it appears that the activity present in the outer membrane fraction is malonyl-CoA sensitive. However, in addition to representing only a small amount of the total CPT activity, when the recovery of the malonyl-CoA sensitive activity in the outer membrane is calculated, it is similar to that observed with cytochrome C oxidase. In summary, the data suggest that the small amount of CPT activity recovered in the outer membrane fraction represents contamination from the inner membrane.

TABLE 3.

Carnitine palmitoyltransferase: specific activity (mU/mg protein) in fractions obtained with French Press treatment of rat liver mitochrondria.

	CPT-Forward Reaction	Malonyl-CoA Sensitive CPT
	(mU/mg protein)	
Inner membrane matrix	115.6	0.8
Supernatant	4.5	0.8
Outer membrane	1.9	1.8

The scorecard for the different approaches to removing the outer membrane gives 3 procedures, that yield results showing that CPT is an inner membrane enzyme, while one procedure discloses in the presence of malonyl-CoA sensitive CPT activity present in the outer membrane. With the phosphate-swelling procedure, the data also suggest that the recovery of CPT activity in the outer membrane is not concordant with the recovery of monoamine oxidase, the outer membrane marker.

C. DIGITONIN METHOD FOR MITOCHONDRIAL FRACTIONATION

It seems worthwhile to review briefly the basis for the digitonin method for mitochondrial membrane fractionation. Digitonin produces the release of adenylate kinase (an intermembrane marker) and monoamine oxidase from mitochondria in a concentration dependent manner.[8] The protein concentration used is also critical; the effect of digitonin is different when the mitochondrial protein concentration is 50 mg/mL versus 5 mg/mL. In addition, the ratio of digitonin to mitochondrial protein is critical in these experiments. Why does digitonin disrupt and dislodge the outer membrane from the inner membrane? The data of Kottke et al,[22] suggests that the cholesterol content of rat brain mitochondrial membranes is different. They observed that the inner membrane contained only 10% the amount of cholesterol compared to the outer membrane. Furthermore, the cholesterol content of the contact sites between the inner and outer membrane were somewhat in between the separated membranes.

In a series of elegant studies from the laboratory of Brdiczka[23,24,25,26] a clear picture has emerged concerning the structure and functional consequences of the organization of the outer membrane, the inner boundary membrane, the cristae and the contact sites where the outer and inner boundary membranes are fused. A review focus on this subject.[27] A brief overview of the findings from these studies might help in attempting to resolve the discordant data concerning CPT and the outer membrane. Digitonin treatment of mitochondria leads to a concentration dependent release of porin, an outer membrane protein. At the same time, hexokinase (type 1) remains bound to the inner membrane-matrix fraction.[24,28] These data are interesting because porin is supposed to be the binding site for hexokinase on the outer membrane and porin is randomly distributed throughout the outer membrane. Furthermore, glutathione S-transferase activity also remains bound to the inner membrane matrix fraction and this enzyme is an outer membrane enzyme. The octameric form of creatine kinase (Mi-CK) is resistant to extraction by digitonin from the inner membrane-matrix and this enzyme is usually considered to be localized within the intermembrane space.

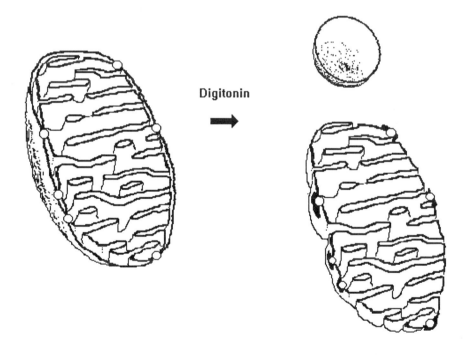

Figure 2. A pictoral representation of the action of digitonin on mitochondria. The contact areas are shown as open circles where the outer membrane and inner boundary membrane are fused.

When the outer membrane fraction obtained by phosphate swelling of mitochondria is subjected to sucrose gradient centrifugation, in addition to a peak representing the outer membrane, there is a section of the gradient that has some very interesting properties. This area has been designated as containing the contact sites or those sections where the outer membrane is fused with the inner boundary membrane.[24] Previous ultrastructural studies had shown the existence of these contact sites.[29,30] The contact sites have a 3-5 times higher binding capacity for hexokinase compared to the outer membrane fraction.[24,31] The contact sites contain glutathione S-transferase activity that has a specific activity about 40 times that observed in the outer membrane.[24] While the activity in the outer membrane fraction can be activated by preincubation with NEM, there is no activation of the activity in the contact sites. The contact sites have glycerol-3-phosphate (FAD) dehydrogenase activity, an enzyme localized to the inner boundary membrane[23] and they also contain creatine kinase (Mi-CK).[25,26,32]

A conceptual diagram (Figure 2) of the action of digitonin is shown above. The outer membrane around the contact sites is disrupted and dislodged with the contact sites remaining with the inner membrane-matrix fraction that is devoid of most of the outer membrane. Within the contact sites, porin and enzymes such as glutathione S-transferase are activated in such a fashion that their functional properties are altered. The mechanism responsible for this activation is unknown, but it is interesting to speculate that it may be a reversible covalent modification of the proteins. Why this should occur only at the contact sites awaits further characterization and study of these interesting submitochondrial particles. In the release of the outer membrane from the mitochondria with digitonin, the inner membrane-matrix fraction undergoes distortion and modification with structural features such as protruding, finger-like areas. The simplified diagram that is portrayed here does not show these distortions.

The diagram shown in Figure 3 demonstrates that during the French Press procedure, the mitochondria are contracted under hyperosmotic conditions. Following disruption of the outermembrane, the resultant inner membrane-matrix fraction retains the normal structure but has obvious areas where there are small tags of outer membrane still attached (the contact sites). Thus, both the digitonin and French Press procedures produce similar types of fractions. The resultant outer membrane does not contain CPT activity, but this activity is retained within the inner membrane-matrix fraction.

When mitochondria are subjected to the phosphate-swelling and shrinkage membrane isolation procedure, there is a greater amount of CPT activity which migrates with the crude outer membrane fraction when compared to the digitonin and French Press methods. As shown in the cartoon (Figure 4) depicting the swelling/shrinkage procedure, both outer membrane and contact sites are released.[24] It is possible that the CPT activity that is attributed to the outer membrane represents activity present in the contact sites.

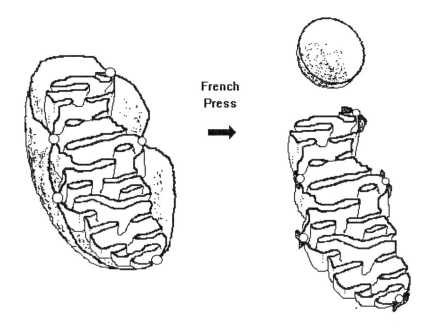

Figure 3. A pictoral representation of the effect of the shearing force generated by the French Press. As in Figure 2., the contact sites are depicted as open circles.

In preliminary experiments, we have been able to show that CPT activity is present within the contact sites and that the specific activity is greater than that observed in the outer membrane peak. However, the situation seems to be more complicated than just this explanation. We suggest that during swelling/shrinkage there is removal of CPT from the inner boundary membrane associated with the contact sites with a redistribution such that now activity is also present within the outer membrane vesicles.

In summary, the data obtained using the French Press to produce a shear force to fracture the outer membrane of rat liver mitochondria is consistent with the data obtained using the digitonin method. These data show that CPT is not an outer membrane enzyme but is retained in the inner membrane-matrix fraction. However, CPT specific activity appears to be enriched in the contact sites following swelling/shrinkage. This suggests that CPT is localized on the inner boundary membrane component of the contact site. It is tempting to speculate that delineation of the organization of the cascade of reactions that lead to the movement of a long-chain acyl group through the mitochondrial membranes and into the matrix may be intimately related to the structure of the contract sites.

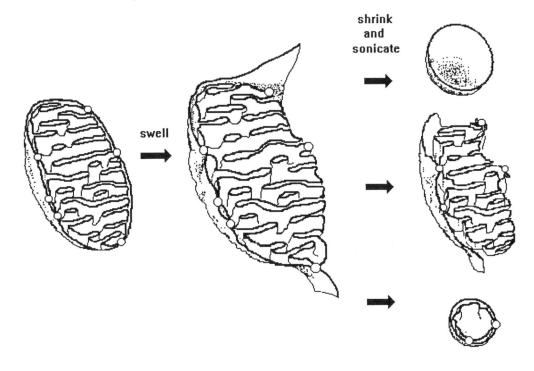

Figure 4. A pictoral representation of the effect of swelling followed by shrinkage with sonication on the release of outer membrane and contact sites from mitochondria. The open circles represent the contact sites between the membranes.

ACKNOWLEDGMENTS: I thank Julia Turkaly and Lauri Albers for excellent technical assistance, Peter Turkaly for the art work in Figures 2-4, and Paul Minkler and Steve Ingalls for help with the manuscript. The work reported here was supported by the Department of Veterans Affairs Medical Research Service and Maternal and Child Health grant # MCJ-009122-03.

REFERENCES:

1. Bremer, J., Carnitine - metabolism and function, <u>Phys. Rev</u>., 63, 1420, 1983.

2. Hoppel, C.L. and Brady, L., Carnitine palmitoyltransferase and transport of fatty acids in <u>The Enzymes of Biological Membranes, vol 2.</u>, Martonosi, A.N., Ed., Plenum Publishing Corp., New York, 1985. 139.

3. Bieber, L.L., Carnitine, <u>Ann. Rev. Biochem</u>., 57, 261, 1988.

4. McGarry, J.D., Woeltje, K.F., Kuwajima, M., and Foster, D.W., Regulation of ketogenesis and the renaissance of carnitine palmitoyltransferase, <u>Diabetes/Metabolism Rev.</u>, 5, 271, 1989.

5. Zierz,S., Engel, A.G., Are there two forms of carnitine palmitoyltransferase in muscle?, <u>Neurology</u>, 37, 1785, 1987.

6.   Long C.S., Haller,R.G., Foster, D.W., and McGarry, J.D., Kinetics of carnitine dependent fatty acid oxidation: Implications for human carnitine deficiency, Neurology, 32, 663, 1982.

7.   Wood, J. McM., Effect of ionic strength on the activity of carnitine palmityltransferase 1, Biochemistry, 12, 5268, 1973.

8.   Hoppel, C.L. and Tomec, R.J., Carnitine palmityltransferase: location of two enzymatic activities in rat liver mitochondria, J. Biol. Chem., 247, 832, 1972

9.   Trevisan, C.P., Angelini, C., Freddo, L., Isaya, G., and Martinuzzi, A., Myoglobinuria and carnitine palmitoyltransferase (CPT) deficiency: studies with malonyl-CoA suggest absence of only CPT-II, Neurology, 34, 353, 1984.

10.   Demaugre, F., Bonnefont, J.-P., Cepanec, C., Scholte, J., Saudubray, J.-M., and Leroux, J.-P., Immunoquantitative analysis of human carnitine palmitoytransferase I and II defects, Ped. Res., 27, 497, 1990.

11.   Brosnan, J.T., Kopec, B., and Fritz, I.B., The localization of carnitine palmitoyltransferase on the inner membrane of bovine liver mitochondria, J. Biol. Chem., 248, 4075, 1973.

12.   Parsons, D.F., Williams, G.R., and Chance, B., Characteristics of isolated and purified preparations of the outer and inner membranes of mitochondria, Ann. N.Y. Acad. Sci., 137, 643, 1966.

13.   Sottocasa, G.L., Kuylenstierna, B., Ernster, L., and Bergstrand, A., An electron-transport system associated with the outer membrane of liver mitochondria, J. Cell Biol., 32, 415, 1967.

14.   Werner, S., and Neupert, W., Functional and bologenetic heterogeneity of the inner membrane of rat-liver mitochondria, Eur. J. Biochem.,25, 379, 1972.

15.   Murthy, M.S.R. and Pande, S., Malonyl-CoA binding site and overt carnitine palmitoyltransferase activity reside on the opposite sides of the outer mitochondrial membrane, Proc. Natl. Acad. Sci. U.S.A., 84, 378, 1987.

16.   Murthy, M.S.R. and Pande, S., Some differences in the properties of carnitine palmitoyltransferase activities of the mitochondrial outer and inner membranes, Biochem. J., 248, 727, 1987.

17.   Murthy, M.S.R. and Pande, S., Characterization of solubilized malonyl-CoA-sensitive carnitine palmitoyltransferase from the mitochondrial outer membrane as a protein distinct from the malonyl-CoA-insensitive carnitine palmitoyltransferase of the inner membrane, Biochem. J., 268, 599, 1990.

18.   Decker, G.L., and Greenawalt, J.W., Ultrastructural and biochemical studies of mitoplasts and outer membranes derived from french-pressed mitochondria, J. Ultrastruct. Res., 59, 44, 1977.

19.   Greenawalt, J.W., Survey and update of outer and inner mitochondrial membrane separation, Methods in Enzymology, 55, 88, 1979.

20. Lilly, K., Burgaisky, G.E., Umeda, P. K., and Bieber, L.L., The medium-chain carnitine acyltransferase activity associated with rat liver microsomes is malonyl-CoA sensitive, Arch. Biochem. Biophys., 280, 167, 1990.

21. Derrick, J.P. and Ramsay, R.R., L-Carnitine acyltransferase in intact peroxisomes is inhibited by malonyl-CoA, Biochem. J., 262, 801, 1989.

22. Kottke, M., Adam, V., Riesinger, I., Bremm, G., Bosch, W., Brdiczka, D., Sandri, G., and Panfili, E., Mitochondrial boundary membrane contact sites in brain: points of hexokinase and creatine kinase location, and control of Ca^{2+} transport, Biochim. Biophys. Acta, 935, 87, 1988.

23. Brdiczka, D., Dolken, G., Krebs, W., and Hofmann, D., The inner boundary membrane of mitochondria, Ho.-Se. Z. Phys. Chem., 355, 731, 1974.

24. Ohlendieck, K., Riesenger, I., Adams, V., Krause, J. and Brdiczak, D., Enrichment and biochemical characterization of boundary membrane contact sites from rat-liver mitochondria, Biochim. Biophys. Acta, 860, 672, 1986.

25. Adams, V., Bosch, W., Schlegel, J., Wallimann,. T., and Brdiczka, D., Further characterization of contact sites from mitochondria of different tissues: topology of peripheral kinases, Biochim. Biophys. Acta, 981, 213, 1989.

26. Kottke, M., Adams, V., Wallimann, T., Nalam, V.K., and Brdiczka, D., Location and regulation of octameric mitochondrial creatine kinase in contact sites, Biochim. Biophys. Acta, 1061, 215, 1991.

27. Brdiczka, D., Knoll, G., Riesinger,I., Weiler, U., Klug, G., Benz, R., and Krause, J., Microcompartmentation at the mitochondrial surface: its function in metabolic regulation in Myocardial and Skeletal Muscle Bioenerfetics, Advances in Experimental Medicine and Biology, vol 194, Brautbar, N., Ed., Plenum Press Corp., New York, 1984.

28. Dorbani, L., Jancsik, V., Linden, M., Leterrier, J.F., Nelson, B.D., and Rendon, A., Subfractionation of the outer membrane of rat brain mitochondria: evidence for the existence of a domain containing the porin-hexokinase complex, Arch. Biochem. Biophys., 252, 188, 1987.

29. Andrews, P.M., and Hackenbrock, C.R., A scanning and stereographic ultrastructure analysis of the isolated inner mitochondrial membrane during change in metabolic activity, Exper. Cell Res., 90, 127, 1975.

30. Hackenbrock, C.R. and Miller, K.J., The distribution of anionic sites on the surfaces of mitochondrial membranes, J. Cell Biol., 65, 615, 1975.

31. Ardail, D., Louisot, P., and Levrat, C., Characterization of the submitochondrial compartments: study of the site of synthesis of dolichol and dolichol-linked sugars, Biochem. Biophys. Res. Comm., 164, 1009, 1989.

32. Biermans, W., Bernaert, I., De Bie, M., Nijs, B., and Jacob, W., Ultrastructure localization of creatine kinase activity in the contact sites between inner and outer mitochondrial membranes of rat myocardium, Biochim. Biophys. Acta, 974, 74, 1989.

CARNITINE PALMITOYL TRANSFERASES: HOW MANY AND HOW TO DISCRIMINATE?

S.V. Pande, A.K.M.J. Bhuiyan, and M.S.R. Murthy*

From the Clinical Research Institute of Montreal affiliated to l'Universite de Montreal

The question of how many carnitine palmitoyltransferases (CPT) isoenzymes there are and how to discriminate them continues to remain unanswered. We provide here our perspective on these issues with some previously unpublished data on the peroxisomal carnitine palmitoyltransferase activities.

ON NOMENCLATURE: CPT REFERS TO CARNITINE LONG-/MEDIUM—CHAIN ACYLTRANSFERASE

Sometimes the same carnitine acyltransferase activity is referred as CPT, COT (carnitine octanoyltransferase), or CDT (carnitine decanoyltransferase). This creates some ambiguity. When the name COT or CDT is used it does not immediately convey whether the medium-chain acyltransferase activity being followed is due to the action of carnitine acetyltransferase, CPT, or both. A carnitine acyltransferase that shows activity exclusively with medium-chain acyl esters of CoA and carnitine has not unequivocally been identified. The various CPT preparations that are active with palmitoyl—CoA as a substrate are able to use medium-chain-acyl esters also and usually with higher reaction velocities under the commonly employed assay conditions. Referring to these activities as carnitine long-chain / medium-chain acyltransferase(s) would be preferable for clarity. The kinetic criterion of letting the experimentally determined values of $V_{max}/K_{0.5}$ with different acyl—CoA esters help identify the preferred acyl—CoA substrate for catalysis and to name the carnitine acyltransferase accordingly would not eliminate the ambiguity completely. This is because the activity and the kinetic parameters, including malonyl—CoA inhibition of these enzymes, substrates of which are amphipathic, are influenced by numerous factors of the assay system like salt, pH, temperature, total protein concentration, bovine serum albumin, phospholipids, thiol reagents, agents causing membrane perturbations, etc. Accordingly, variable preferences are likely to be picked up for different chain length acyl—CoA substrates under different assay conditions. Presently, we are referring to these long- / medium-chain acyltransferase activities by the more commonly used and understood abbreviation of CPT.

*Present address: Department of Biochemistry, Michigan State University, East Lansing, MI 48824, USA

I. CPT ACTIVITIES RESIDE IN VARIOUS INTRACELLULAR LOCATIONS

CPT, the carnitine long-chain acyl transferase activity was believed not too long ago to be an exclusively mitochondrial activity. We now know with fair certainty that the carnitine long-/medium-chain acyltransferase activity is localized also in microsomes and peroxisomes. The mitochondrial CPT activity is found in two locations; the malonyl—CoA sensitive, detergent labile, CPT_0 activity resides in the outer membrane and the malonyl—CoA insensitive, relatively detergent stable, CPT_i activity resides in the inner mitochondrial membrane[1].

Occurrence of a true soluble CPT activity in the "cytosolic" fractions has not been described. The only long-/medium-chain carnitine acyltransferases activity, referred as COT or CPT, that readily appears as "soluble" on manipulations in vitro is believed to be derived from the matrix of peroxisomes. All other CPT activities are found associated with membranes. This association is loose for the inner CPT of the inner mitochondrial membrane but strong for the malonyl—CoA inhibitable CPT of the outer mitochondrial, the microsomal, and the peroxisomal membranes.

A membrane bound form of CPT activity is present in the cell membranes of erythrocytes[2,3]. Not much is known about this enzyme system, however.

We may now consider the question of whether the CPT activities in various intracellular locations result from the activity of different catalytic CPT proteins and, if so, how many. In covering the necessary evidences, we will go through the approaches that have proven helpful in the discrimination of the different CPT activities.

A. MITOCHONDRIAL CARNITINE PALMITOYLTRANSFERASES

Whether the outer and the inner carnitine palmitoyltransferase activities of mitochondria are expressions of the same or different proteins continues to remain unresolved because of the apparently divergent results being obtained in different laboratories. We describe here evidences obtained in our and other laboratories that, we believe, show that different catalytic proteins are responsible for the CPT_0 and CPT_i activities.

Using malonyl—CoA sensitivity as the marker of the outer membrane CPT, we have found considerable differences between the properties of the malonyl—CoA sensitive, CPT_0, activity of the outer membrane with the malonyl—CoA insensitive, CPT_i, activity of the inner membrane4. However, the possibility that such differences arise mostly from differences in the membrane environment of the two differently placed CPT activities cannot be eliminated. Evidences obtained using approaches that are free from the limitation of membrane components affecting the results have clearly shown however that the catalytic activities of CPT_o and CPT_i reside in different proteins. These are as follows.

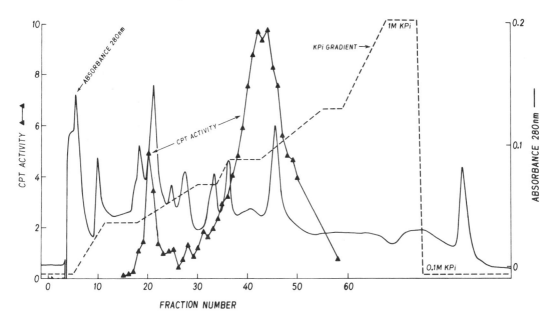

Fig.1 Fractionation of liver mitochondrial outer membrane proteins and carnitine palmitoyltransferase activities by HPLC on a hydroxylapatite column. Rat liver outer membrane vesicles were solubilized using octylglucoside in presence of glycerol. About 1 mg of the solubilized protein was injected into a hydroxylapatite column (from Toyosoda, Tokyo, Japan) and a potassium phosphate (KPi) gradient was used for their elution. CPT activity was monitored with 50 μM [^{14}C]octanoyl-CoA and 10 mM carnitine. Other details were as in Murthy and Pande[5].

(1) When the inner membrane vesicles of liver mitochondria are solubilized with octylglucoside and the proteins are fractionated by HPLC on a hydroxylapatite column using a phosphate gradient, the CPT_i activity appears as a single peak near 260 mM potassium phosphate. Under the same conditions with the outer membrane vesicles only a minor CPT activity appears in this region; most of the CPT activity in this case appears instead at a potassium phosphate concentration of near 500 mM (peak II)[5].

(2) The peak II CPT activity derived from outer membrane vesicles rapidly loses activity. The CPT activity of peak I on the other hand, whether derived from inner membrane vesicles or the outer membrane vesicles, is relatively stable.

(3) The peak II activity shows marked inhibition by low micromolar concentrations of malonyl—CoA. On reconstitution into

asolectin liposomes, the peak II activity shows increase in assayable CPT activity as well as an increase in its malonyl–CoA inhibition. The peak I CPT activity does not share these properties.

(4) A polyclonal antibody raised against homogeneous CPT_i preparation purified using rat liver inner membrane vesicles as the starting material precipitates the peak I CPT activity but not that of the peak II. In Western blots, this antibody reacts positively with a single polypeptide of ca. 69 kDA with the purified CPT_i preparations as well as with peak I preparations; conversely, it shows no reactivity with peak II preparations having malonyl–CoA sensitive CPT activity[5]. These results are consistent with the findings from other laboratories that immunologically the inner and the outer CPT activities are distinct proteins[6-8]. Along the same lines, Singh and associates have reported that in Western blots, using antibody to the detergent stable form of CPT, likely CPT_i, whereas muscle preparations from "normals" showed a 69 kDA reacting band. This antigen band was absent in a corresponding preparation from the muscular form of CPT deficiency patient[9]. Similarly, Saudubray's group has shown that in immunoquantitation experiments, the antibody to the inner CPT showed decreased quantities of a cross reacting material in fibroblasts from muscular form of CPT deficiency patients; this was not the case in preparations from the patients with the hepatic form of the carnitne palmitoyltransferase deficiency disease[10-12]

1. Approaches Used/Tried for the Selective Measurement of the Outer and the Inner CPT Activities of Mitochondria

Certain studies particularly those dealing with the carnitine palmitoyltransferase deficiency disease require discrimination of these two mitochondrial CPT activities. Inasmuch as these two activities reside in immunologically distinct proteins, application of immunoquantitation approaches seem to be of choice. However, other considerations, including the possibility that CPT deficiency may, at times, result from the fabrication of a mutant protein instead of from the lack of a particular CPT protein, require availability of functional assay procedures that would be applicable for the discrimination of the various CPT activities in tissue extracts. Many such approaches have been tried with varying degree of success. We briefly outline here our assessment of the reliability and limitations of these procedures believing that this would aid in the design of appropriate experiments and in the proper interpretation of some of the related existing data.

The belief that the selective assay of the outer CPT activity is made possible by employing low palmitoyl-CoA and carnitine concentrations, while only the inner CPT activity is picked up in a assay that contains a high (0.7 mM) palmitoyl [^{3}H]carnitine concentration, has not been found helpful[13,14].

The possibility that the selective assay of the inner CPT should be possible following inactivation of the outer CPT by tetradecylglycidyl-CoA has been examined. It was not found suitable, however, because the near 20 μM tetradecylglycidyl-CoA concentration required for the full inhibition of the outer CPT was found to partially inhibit also the inner CPT[14]. Attempts to exploit the reported ability of the inner CPT to use tetradecylglycidyl-CoA itself as the carnitine acceptor in the CPT$_i$ assay also

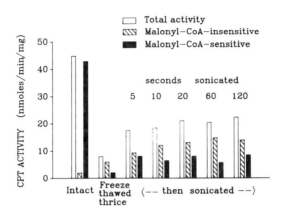

Fig.2 Freeze-thawing of skeletal muscle mitochondria decreased (masked) their CPT activity that was partially restored (unmasked) by subsequent sonication. Mitochondrial suspension (1 mg/ml) in isotonic sucrose-mannitol-Tris medium were allowed to freeze completely (solid CO_2) then to thaw (at 30°C until just melted). This freeze-thaw process was repeated thrice. CPT activity was then monitored before and after sonication. From Pande et al 1990. Reproduced by permission of the Biochim Biophys. Acta, vol. 1044, p. 262, 1990.

proved futile because the inhibition of the CPT$_i$ activity observed with tetradecylglycidyl-CoA as a substrate was found to increase with time[14,15].

Another approach has been to measure the CPT activity with preparations having largely intact mitochondria and to consider the assayable activity to be due to the outer CPT. The same activity is then measured following disruption of mitochondria by freeze-thaw / sonication or by detergents and the resultant increase in activity is ascribed as being due to the inner CPT. We have found, however, that freeze-thaw and sonication are not reliable in this regard because these procedures cause a masking of the outer CPT activity while exposing parts of the inner CPT activity (Figure 2);

formation of membrane vescicles having mixed membrane orientation appear to be responsible for this phenomenon[14]. We have subsequently observed that the freeze-thaw of skeletal muscle tissue as well as of their homogenates also leads to the partial masking of their malonyl—CoA sensitive CPT activity while some of their malonyl—CoA insensitive activity is concurrently exposed.

The realization that the outer mitochondrial CPT activity is detergent labile while the inner CPT activity is relatively stable has led to the use of detergent exposure of mitochondrial preparations to help discriminate between the CPT_i and the CPT_0 deficiencies forms[16]; however, the exact conditions of detergent exposure have varied in these experiments and their reliability has not been critically examined. We have found using outer and the inner membrane preparations of liver mitochondria as the respective sources of the outer and the inner CPT activities that the 1 mM concentration of palmitoyl(+)carnitine does not inactivate the outer CPT activity even when incubations and prior incubations are performed at 37°C. A selective inactivation of the outer CPT activity does result with high, 50 to 150 μM, concentrations of palmitoyl-CoA, but only when ug quantities of membrane proteins are present as the source of CPT. However, even with 150 μM palmitoyl—CoA, some, about 10%, of the CPT activity of outer membrane vesicles survives while with inner membrane vesicles a partial inhibition of the inner CPT activity becomes manifest under these conditions. Despite these reservations, a careful use of detergents seems to provide a means of discriminating between CPT_i and CPT_0 activities. In tissue extracts the contribution of nonmitochondrial detergent labile and detergent insensitive CPT activities to the mitochondrial CPT_0 and CPT_i values also need to be taken into account. However, in muscles and also in fibroblasts, it is likely that the contribution of the nonmitochondrial CPT to the total CPT values would be minimal.

A useful method for discrimination between CPT_0 and CPT_i has been to regard the malonyl-CoA inhibitable CPT activity to represent the outer CPT activity while to consider the malonyl—CoA insensitive CPT activity to be due to the inner CPT. Malonyl—CoA inhibition is, however, remarkably affected by the assay conditions like temperature, pH, BSA, acyl-CoA concentration, the amount of mitochondrial protein present during the assay, preparation of muscle mitochondria with proteases, conditions of prior incubation, the concentration of carnitine used, etc.[1,17,18]. We have found that to this known list should be added the need to control the salt concentration; the malonyl-CoA sensitivity of CPT is markedly decreased by increase in salt concentration. This necessitates a careful control of assay conditions as failure to do so can lead to erroneous conclusions. For example, manipulations that increase the salt concentrations of the assay medium, such as use of the fractions eluted from various columns with high salts, frequently lead to a decrease in the malonyl—CoA inhibition of the CPT_0 activity. And this can readily be misinterpreted as a conversion of the malonyl—CoA sensitive CPTo to the insensitive form by the fractionation procedure involved.

An exact assessment of the CPT_0 and CPT_i activities of tissues by monitoring the malonyl—CoA inhibition of the total CPT activity and regarding the malonyl—CoA sensitive fraction as representing the CPT_0 and the insensitive fraction as the CPT_i is also only of limited value. This is because procedures that would allow full exposure of both the outer and the inner CPT activities at the same time without accompanying undesirable side effects are not available. As mentioned above freeze-thaw is unsuitable in this regard because of its masking and unmasking effect of the two CPT activities. Similarly, use of detergents are not useful in this setting because most of them cause activation of CPT_i activity, inactivation of CPT_0 activity and a reduction of the malonyl—CoA sensitivity of the CPT_0[4,6].

The observations that only the malonyl-CoA sensitive CPT of both mitochondria and peroxisomes show activity with aminocarnitine and octanoyl-CoA as substrates whereas the malonyl-CoA insensitive isozymes of both these org/anelles do not show this activity[19] initially seemed to be of some promise for the discrimination of CPT activities. However, this has not been the case because the activities realized with aminocarnitine and octanoyl-CoA as substrates are rather too low to allow a good signal to noise difference needed for reliable estimates of the activities. Besides, it is not known in this regard whether the microsomal malonyl-CoA sensitive CPT would behave similar to the malonyl—CoA sensitive CPT activities of the outer mitochondrial and the peroxisomal membranes.

2. Mitochondrial CPT_0 Isozymes?

Indirect evidences indicate that the mitochondrial malonyl—CoA sensitive outer CPT exists in two tissue type isozymic forms, the liver type and the muscle type. When examined in their membrane bound forms these two forms show differences in their kinetic properties and in their sensitivity to inhibiton by malonyl—CoA and other inhibitors[20]. McGarry and associates have shown that with mitochondria of muscle, tetradecylglycidyl—CoA labeling yields a peptide of slightly smaller mol wt, of about 86 kDa than, with liver mitochondria, about 94 kDa[21]. We have obtained similar results on labeling with etomoxir in presence of ATP and CoA[5]. As these CPT_0 activities have not been sufficiently characterized, it is not known how to discriminate them. Clearly, labeling with tetradecylglycidyl—CoA or etomoxir is unsuitable for the routine discrimination of these two isozymes. Besides, the question of whether the radiolabeled bands so identified represent the regulatory or the catalytic peptides of the CPT_0 system remains to be clarified as does the question of whether these two functions indeed reside in separate peptides as some investigators believe. The possibility of a regulatory, malonyl—CoA binding, peptide being separate from the catalytic peptide of the CPT_0 has been described[22]. However, such experiments have been carried out and interpreted on the basis of the belief that the same peptide is responsible for the catalytic function of both the CPT_i and CPT_o[23-25]. This latter, however, is inconsistent with bulk of the other evidences as described above.

B. MICROSOMAL CARNITINE PALMITOYLTRANSFERASE

A malonyl–CoA sensitive CPT is present in microsomes presumably in its membrane. This activity is labile to Triton and to high concentrations of palmitoyl–CoA. Antigenically, this activity seems different from both the detergent stable CPT (i.e. CPT_i) of mitochondria and the readily solubilizable, malonyl–CoA insensitive CPT activity of peroxisomes[26]. We are not aware of any specific method that may allow discrimination of the microsomal, malonyl–CoA sensitive CPT activity from the other malonyl–CoA sensitive activities of tissue extracts.

C. PEROXISOMAL CARNITINE PALMITOYLTRANSFERASE(S)

Peroxisomes have been known to have a readily solubilizable COT (CPT) activity that purifies as a monomer of ca. 63 kDa. This enzyme is relatively insensitive to malonyl–CoA inhibition. Immunologically, this activity is distinct from the CPT_i of mitochondria[27]. Because of its monomeric nature, molecular sieving allows the separation of this activity from the other membrane associated CPT activities in liver[28]. However, this approach is not suitable for the routine quantitation of the peroxisomal soluble CPT activity in crude liver extracts.

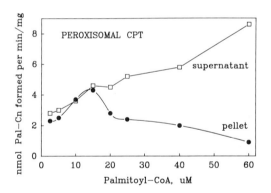

Figure 3. Differential effects of palmitoyl–CoA on the readily solubilizable and the membrane bound CPT activities of peroxisomes. The two CPT preparations were obtained as described in Table 1. The CPT activity was measured with [^{14}C]-palmitoyl-CoA as in 4.

In gradient purified peroxisomes the membrane associated CPT/COT activity shows a number of differences in its properties from those seen with the readily solubilizable form of CPT mentioned above. Thus, Derrick and Ramsay showed that the CPT/ COT activity of whole peroxisomes and of a pellet fraction obtained following sonication and centrifugation of peroxisomes is as sensitive to malonyl—CoA inhibition as the mitochondrial CPT_0[29]. The chain length activity profile of the carnitine acyltransferase in whole peroxisomes was also found to match that observed with intact mitochondrdia. The CPT activity of whole peroxisomes or the membrane pellet derived from them shows additional similarities to the mitochondrial outer CPT and a marked differences to the readily solubilizable form of peroxisomal CPT. Thus, the CPT activity of whole peroxisomes measured with palmitoyl—CoA and carnitine is inhibited by l-aminocarnitine like that of the mitochondrial CPT but the corresponding activity of the purified, readily solubilizable form of the peroxisomal CPT is not inhibited[19]. Figure 3 shows that the CPT activity of the peroxisomal pellet was sensitive to inhibition by high concentration of palmitoyl-CoA while that of the readily solubilizable peroxisomal CPT was not. Thus, the malonyl—CoA sensitive CPT of peroxisomal membrane behaves similar to the malonyl—CoA sensitive CPT activity of the outer mitochondrial membrane and of the microsomes[4,26]. Whether the same protein would be responsible for the catalytic and the regulatory properties of the malonyl—CoA senitive CPT activities of these different intracellular membranes remains to be examined. We do know meanwhile that the fatty acyl—CoA ligase activity is also found associated with the outer mitochondrial membrane, the microsomal

TABLE 1

Intraperoxisomal Distribution of CPT Activity

Gradient purified peroxisomes were suspended in 10 mM potassium phosphate (pH 7.4), 1 mM EDTA, 1 mM DTT, 0.8 mM PMSF to about 2 mg protein/ml. Five ml suspensions were sonicated, 3 x 15 sec, 0°C, with 30 sec cooling, then centrifuged, at 250,000 x g for 60 min, to obtain the 1st supernatant. The residue was reprocessed as above to obtain a 2nd supernatant, a brown membrane fraction and a transparent jelly like pellet. CPT activity was measured at 40 μM [^{14}C]-palmitoyl-CoA without or with 100 μM malonyl-CoA present.

	Whole peroxisomes	"Supernatant" 1st	2nd	Brown membrane fraction
CPT activity (palmitoylcarnitine nmol/min/mg)	2.9	3.2	2.8	4.7
Malonyl-CoA inhibition %	88	20	36	82
Recovery protein %	100	10	7	45
CPT units %	100	9	7	73

membrane and the peroxisomal membrane and that the same protein is responsible for the acyl–CoA ligase activities of these different subcellular fractions[30].

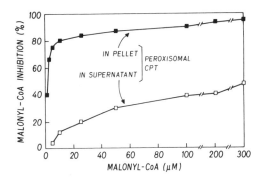

Figure 4. Differences in the malonyl-CoA sensitivity of the readily solubilizable and the membrane associated CPT activity of peroxisomes. Assays were with 40 uM palmitoyl-CoA and 1.6 mg/ml BSA4

In agreement with Derrick and Ramsay 29, we have found that from the gradient purified peroxisomes of liver, when assayed with palmitoyl–CoA as a substrate, the readily solubilizable CPT activity released in the supernatant represented a minor fraction of the initial total CPT activity and that it was relatively malonyl–CoA insensitive (Table 1 and Figure 4).
The major part of the CPT, about 2/3rd of the total, was found associated with the membrane fraction and it showed marked malonyl-CoA sensitivity (Table 1 and

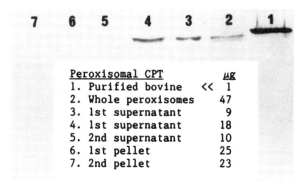

Figure 5. Differences in the immunoreactivity of the readily solubilizable and the membrane bound CPT activities of peroxisomes of rat liver. The proteins of the different peroxisomal fractions, obtained as in TABLE 1, were precipitated (except purified one) with 20% trichloroacetic acid, subjected to SDS-PAGE, transferred to polyvinylidine difluoride membrane, and then probed with the anti-bovine CPT antibody to the peroxisomal soluble CPT.

Figure 4). Western blots using the antibody to the readily solubilizable CPT of bovine peroxisomal origin showed that the cross reacting antigen appeared in the first supernatant of the peroxisomes with barely detectable amounts in the second supernatant and none in the peroxisomal pellet (Figure 5). This was despite the fact that in the experiment of Figure 5, in terms of the CPT units, lanes 6 and 7 had over three times as many enzyme units as those in lane 3. These findings clearly show that the malonyl–CoA sensitive CPT of peroxisomal membrane and the readily solubilizable form of peroxisomal CPT that is malonyl–CoA insensitive are distinct proteins.

II. CONCLUSIONS

We believe the information summarized above allows the following general conclusions to be drawn:

(a) That there are two readily identifiable malonyl–CoA insensitive CPT activities in liver; one resides in the inner mitochondrial membrane and the other in the peroxisomal matrix. These are distinct proteins.
(b) That the malonyl–CoA sensitive CPT activities are present, in liver at least in three locations; in membranes of mitochondria, microsomes and peroxisomes. Whether these share the same catalytic and regulatory site(s) remains unknown.
(c) That the catalytic peptide of the CPT_o is different from that of the CPT_i.
(d) That the readily solubilizable, malonyl–CoA insensitive CPT of peroxisomes and the malonyl–CoA sensitive membrane bound CPT of peroxisomes are distinct proteins.

It is clear that methods for the unequivocal discrimination of the various CPT activities in tissue extracts are not available. Some group discrimination is possible based on differences in the malonyl–CoA sensitivities, detergent susceptibilities, and immunoreactivites. Availability of monospecific antibodies to the malonyl–CoA sensitive CPT activit(y)ies would be helpful but these have not yet been obtained.

III. REFERENCES

1. Murthy, M.S.R. and Pande, S.V., Malonyl-CoA binding site and the overt carnitine palmitoyltransferase activity reside on the opposite sides of the outer mitochondrial membrane, *Proc. Natl. Acad. Sci.*, 84, 378, 1987a.

2. Wittels, B. and Hochstein P., (for CPT in erythrocyte membrane, Ist report) *J. Biol. Chem.*, 242, 126, 1967.

3. Arduini, A., Mancinelli, G. and Ramsay, R.R. Palmitoyl-L-carnitine, a metabolic intermediate of the fatty

acid incorporation pathway in erythrocyte membrane phospholipids, *Biochem. Biophys. Res. Commun.*, 173, 212, 1990.

4. Murthy, M.S.R. and Pande, S.V., Some differences in the properties of carnitine palmitoyltransferase activities of the mitochondrial outer and inner membranes, *Biochem. J.* 248, 727 1987b.

5. Murthy, M.S.R. and Pande, S.V., Characterization of a solubilized malonyl-CoA-sensitive carnitine palmitoyltransferase from the mitochonmdrial outer membrane as a protein distinct from the malonyl-CoA-insensitive carnitine palmitoyltransferase of the inner membrane, *Biochem. J.*, 268, 599, 1990.

6. Woeltje, K.F., Kuwajima, Foster,D.W. and McGarry, D., Characterization of the mitochondrdial carnitine palmitoyltransferase enzyme system. II. use of detergents and antibodies, *J. Biol. Chem.*, 262,9822, 1987.

7. Woeltje, K.F., Esser, V., Weis, B.C., Cox, W.C., Schroeder, J.G., Shyue—Tsong, L., Foster, D.W., and McGarry, D.J., Inter-tissue and inter-species characteristics of the mitochondrial carnitine palmitoyltransferase enzyme system, *J. Biol. Chem.*, 265, 10714, 1990.

8. Thampy, K.G., Characteriztion of mitochondrial carnitine palmitoyltransferases I and II using malonyl—CoA and emeriamine as inhibitors. *FASEB J.*, 5, Abstr 4560, 1991

9. Singh, R., Shepherd, I.M., Derrick, J.P., Ramsay, R.R., Sherratt, H.S.A. and Turnbull, D.M., A case of carnitine palmitoyltransferase II deficiency in human skeletal muscle, *FEBS Letters,* 241, 126, 1988.

10. Demaugre, F., Bonnefont, J.P., Mitchell, G., Hoang, N..N., Pelet A., Rimodi, M., DiDonato, S. and Saudubray J.M., Hepatic and muscular presentation of carnitine palmitoyltransferase deficiency: Two distinct entities, *Ped. Res.*, 24, 308, 1988.

11. Tein, I., Demaugre, F., Bonnefont, J.P. and Saudubray, J.M., Normal muscle CPT1 and CPT2 activities in hepatic presentation patients with CPT1 deficiency in fibroblasts. Tissue specific isoforms of CPT1? *J. Neurol. Sci.*, 92, 229, 1989.

12. Demaugre, F., Bonnefont, J.P., Cepanec, C., Scholte, J., Saudubray, J.M. and Leroux, J.P., Immunoquantitative analysis of human carnitine palmitoyltransferase I and II defects, *Ped. Res.*, 27, 497, 1990.

13. Scholte, H.R., Jennekens, F.G.I. and Bouvy, J.J., Carnitine palmitoyltransferase II deficiency with normal carnitine palmitoyltransferase I in skeletal muscle and leucocytes, *J. Neurol. Sci.*, 40, 39, 1979.

14. Pande, S.V., Lee, T.S. and Murthy, M.S.R., Freeze-thawing causes masking of membrane-bound outer carnitine palmitoyltransferase activity: implications for studies on carnitine palmitoyltransferases deficiency, *Biochim. Biophys. Acta,* 1044, 262, 1990.

15. Kiorpes, T.C., Hoerr, D., Weaner, L. and Tutwiler, G.F., Identification of 2-tetradecylglycidate (methyl palmitate) and its characterization as an irreversible active site-directed inhibitor of carnitine palmitoyltransferase A in isolated rat liver mitochondria, *J. Biol. Chem.*, 259, 9750, 1984.

16. Lund, H., Carnitine palmitoyltransferase: characterization of a labile detergent-extracted malonyl—CoA sensitive enzyme from rat liver mitochondria, *Biochim. Biophys. Acta.*, 918, 67, 1987.

17. Kolodiej, M.P. and Zammit, V.A., Sensitivity of inhibition of rat liver mitochondrial outer membrane carnitine palmitoyltransferase by malonyl-CoA to chemical- and temperature-induced changes in membrane fluidity, *Biochem. J.*, 272, 421, 1990.

18. Bieber, L.L. and Farrell, S., Carnitine acyltransferases, in *The Enzymes,* Vol. 16, 3rd ed., Boyer, P.D., ed., Academic, New York, 1983, 627.

19. Murthy, M.S.R. and Pande, S.V., Acyl-CoA chain length affects the specificity of various carnitine palmitoyltransferases with respect to carnitine analogues, *Biochem. J.*, 267, 273, 1990.

20. Saggerson,D., Carnitine palmitoyltransferase in extrahepatic tissues, *Biochem. Soc. Transc.*, 14, 679, 1986.

21. Declercq, P.E., Falck, J.R., Kuawjima, M., Tyminshi, H., Foster, D.W. and McGarry, J.D., Characterization of the mitochondrial carnitine palmitoyltransferase enzyme system. I. Use of inhibitors, *J. Biol. Chem.*, 262, 9812, 1987.

22. Kerner, J. and Bieber, L.L., Isolation of a malonyl-CoA sensitive CPT/b-oxidation enzyme complex from heart mitochondria, *Biochemistry,* 29, 4326, 1990.

23. Chung, C., Woldergiorgis, G. and Bieber, L.L., Restoration of malonyl-CoA sensitivity to purified rat heart mitochondrial carnitine palmitoyltransferase by addition of protein fraction(s) from an 86 kD malonyl-CoA binding immoaffinity clumn, *FASEB J.*, Abst. # 1290, 1991.

24. Ghadiminejad, I. and Saggerson, E.D., The relationship of rat liver overt carnitine palmitoyltransferase to the mitochondrial malonyl-CoA binding entity and to the latent palmitoyltransferase, *Biochem. J.*, 270, 421, 1990.

25.   Ghadiminejad, I. and Saggerson, E.D., Carnitine palmitoyltransferase (CPT2) from liver mitochondrial inner membrane becomes inhibitable by malonyl—CoA if reconstituted with outer membrane malonyl—CoA binding protein, *FEBS. Lett.*, 269, 406, 1990.

26.   Lilly, K., Bugaisky, G.E., Umeda, P.K. and Bieber, L.L., The medium-chain carnitine acyltransferase activity associated with rat liver microsomes is malonyl-CoA sensitive, *Arch. Biochem. Biophys.*, 280, 167, 1990.

27.   Ramsay, R.R., The soluble carnitine palmitoyltransferase from bovine liver, *Biochem. J.*, 249, 239, 1988.

28.   Healy, M.J., Kerner, J. and Bieber, L.L., Enzyme of carnitine acylation, *Biochem. J.*, 249, 231, 1988.

29.   Derrick J.P. and Ramsay R.R., L.-carnitine acyltransferase in intact peroxisomes is inhibited by malonyl-CoA, *Biochem. J.*, 262, 801, 1989.

30.   Miyazawa, S., Hashimoto, T. and Yokota, S., Identity of long-chain acyl-coenzyme A synthetase of microsomes, mitochondria, and peroxisomes in rat liver, *J. Biochem.*, 98, 723, 1985.

DISCUSSION OF THE PAPER

T. GREWAY *(Nutley, NJ)*: Is it possible, assuming model 1 of Beiber, that CPT, and CPT 2 deficiencies are due to errors in the signal sequences for insertion in the outer membrane into the inner membrane?

PANDE: In as much as strong evidence indicates that the CPT_1 and CPT_2 are different proteins, this possibility seems remote.

G. WOLDEGIORGIS *(Madison, WI)*: How do you explain that adding a malonyl-CoA binding protein fraction (inactive CPT I) restores malonyl-CoA sensitivity to CPT II?

PANDE: We have no explanation for these reported observations of others. To the limited extend we have examined, we have failed to obtain such results.

G. HUG *(Cincinnati, OH)*: You showed 2 bands of CPT II in liver but just 1 band in heart and muscle. Could this explain our finding of 17% liver CPT II, but less than 2% CPT II in heart and muscle of our patient with generalized CPT II deficiency?

PANDE: Both liver and heart CPT II show only one band in immunoblot experiments when probed with an antibody to the inner CPT.

CLINICAL DEFICIENCIES OF CARNITINE PALMITOYL TRANSFERASE

J.P. BONNEFONT[1], F. DEMAUGRE[1], I. TEIN[4], C. CEPANEC[1],
M. BRIVET[2], D. RABIER[3], J.M. SAUDUBRAY[4]

[1] Laboratoire de Biochimie et Unité INSERM 75, Hôpital des Enfants-Malades, Paris.
[2] Laboratoire de biochimie, Hôpital Bicêtre, Le Kremlin Bicêtre.
[3] Laboratoire de Biochimie B, Hôpital des Enfants-Malades, Paris.
[4] Clinique de Génétique Médicale et Unité lNSERM 12, Hôpital des Enfants-Malades.

Mitochondrial fatty acid oxidation is a major source of energy in man. The entry of long-chain fatty acid (LCFA) into mitochondria is governed by carnitine palmitoyl transferase (CPT), (Ec 2, 3, 1, 21). CPT is classically distributed on both the outer (CPT 1) and the inner (CPT 2) mitochondrial membranes. Recent data suggest that CPT 1 and CPT 2 activities would involve two distinct proteins. LCFA oxidation contributes to energy homeostasis especially in the heart, the liver and the skeletal muscle. In the myocardium, LCFA have been shown to be the preferred substrate in resting state. In the liver, oxidation of LCFA produces ketone bodies enhances gluconeogenesis, and therefore contributes to the maintenance of normoglycemia during fasting. In skeletal muscle, oxidation of LCFA not only plays a role in energy homeostasis during fasting, but also is essential to perform prolonged exercise. Therefore, a simultaneous dysfunction of the liver, the heart and the skeletal muscle is usually reported in patients with defects of mitochondrial LCFA oxidation. Conversely, CPT deficiency is not usually known to cause such a multitissular dysfunction. The more commonly described phenotype was referred to as muscular CPT deficiency (1) and another form was referred to as hepatic CPT deficiency (2). A third presentation associating cardiac, hepatic, and muscular injury was recently reported in one case (3).

I. MUSCULAR CPT DEFICIENCY

The muscular presentation of CPT deficiency has been described in about 60 patients since the first report by Di Mauro in 1973 (4). The clinical presentation is stereotyped (Table 1). The patient is most often an adolescent male and experiences acute episodes of rhabdomyolysis after prolonged

TABLE 1

CPT II DEFICIENCIES

CLINICAL SYMPTOMS (LITERATURE)	
ONSET <6 YEARS	10%
>10 Y<15 Y	50%
TRIGGERS LONG PHYSICAL EXERCISE	95%
VIRAL INFECTION	20%
NOTHING	25%
MUSCLE PAIN	90%
CRAMPS	40%
MYOGLOBINURIA	90%
ACUTE RENAL FAILURE	30%
RESPIRATORY FAILURE	20%
CARDIAC SYMPTOMS	<5%
NORMAL MUSCLES BETWEEN ATTACKS	95%
NUMBER OF ATTACKS > 5	50%

BIOLOGICAL DATA (LITERATURE)	
MYOGLOBINURIA	90%
CREATINE KINASE (X 20 TO 100)	70%
LACTIC DEHYDROGENASE (X 5 TO 10)	90%
ALDOLASES (X 5 TO 10)	50%
TRANSAMINASES GOT (X 200)	50%
TRANSAMINASES GPT (X 10 TO 20)	30%
TRIGLYCERIDES > 2 MMOL/L	50%
SERUM CARNITINE NORMAL OR HIGH	90%
MUSCLE CARNITINE NORMAL OR HIGH	82%
FASTING HYPOGLYCEMIA	0%
FASTING CK (X 2 TO 15)	77%
FASTING LOW KETOGENESIS	71%
LCT: ABNORMAL KETOGENESIS	75%
MUSCLE HISTOLOGY: NORMAL	50%
MUSCLE HISTOLOGY: LIPID DROPLETES	35%

exercise or occasionally after prolonged fasting. Spontaneous attacks are possible but muscle cramps are rare. Renal failure can occur during acute attacks. Urinary organic acid profile is not contributive to the diagnosis. Serum and muscle carnitine levels are not usually decreased but lipid accumulation is occasionally demonstrated in muscle biopsies. CPT deficiency is not restricted to skeletal muscle but is also found in all other tissues studied (eg liver, leukocytes, platelets, fibroblasts). Extramuscular consequences of this defect are however minor. Ketone body production is only delayed or mildly insufficient during fasting or long-chain triglyceride loading in the few patients thus studied. Fasting hypoglycemia has not been observed. Similarly, cardiac dysfunction is not a usual feature of this phenotype; mild ECG abnormalities and minor ventricle hypertrophy have been reported only in a few cases. In fibroblasts, LCFA are normally oxidized in most cases studied. The muscular presentation of CPT deficiency is not usually considered as a severe condition but in one case, death after an attack was ascribed to this defect (5).

In most cases, the methodology of CPT assays does not allow a clear discrimination between CPT 1 and CPT 2 activities. In a few patients, however, CPT 1 and CPT 2 were clearly identified (6). CPT 1 activity was normal whereas CPT 2 activity was reduced (residual activity: 25% of control values) (Table 2). In these patients, the enzymatic defect was shown to result from a decreased amount of the protein bearing CPT 2 activity (7). This abnormality correlated with a decreased biosynthesis of CPT 2 in one patient (8). This CPT 2 abnormality was probably ubiquitous since no tissue-specific isoforms of CPT 2 have been demonstrated in human liver, skeletal muscles, fibroblasts, and heart (3). In this way, the lack of extramuscular symptoms in these patients was puzzling.

TABLE 2
CPT Activities in Fibroblasts from Patients with the
"Muscular" and the "Hepatic" Presentation

Malonyl CoA 50 μM		Controls n=12	"Muscular"				"Hepatic"		
			1	2	3	4	1	2	3
CPT I	−	2.4 ± 0.65	3.3	2.8	1.5	3.1	0.39	0.56	0.33
	+	0.23 ± 0.06	0.30	0.22	0.15	0.71	0.21	0.26	0.15
CPT II	−	2.30 ± 0.32	0.60	0.67	0.35	0.31	2.50	2.02	2.40

Results (mean ± SD) are expressed as nmol of palmitoyl L carnitine produced per min per mg NCP.

II. HEPATO-CARDIO-MUSCULAR CPT DEFICIENCY

The hepato-cardio-muscular presentation of CPT deficiency was reported by Demaugre in 1991 (3). A three-month old boy suffered from a severe hypoketotic hypoglycemia with hepatomegaly and liver failure. Heart injury was expressed as heart beat disorders with a mild cardiac enlargement. Skeletal muscle injury was restricted to a mild increase of plasma creatine kinase level. Urine organic acid profile was normal. Plasma total carnitine was 14 μmol/l (normal 28-71), and free carnitine was 3 μmol/l (normal 16-50). Long- and medium-chain triglyceride loadings performed at the age of 6 months indicated the presence of a metabolic block on the pathway of LCFA oxidation in liver (Figure 1). In fibroblasts, palmitate oxidation was decreased whereas octanoate and butyrate were normally oxidized. CPT 1 activity was normal whereas CPT 2 activity was markedly reduced (10% of control values). The defect was shown to result from a decreased amount of the protein bearing CPT 2 activity, as previously shown in fibroblasts from two CPT 2 deficient patients with a muscular form (7). In vivo as in vitro, functional consequences of the CPT 2 defect were not identical in both sets of patients. Hepatic and cardiac injury, as well as impaired LCFA oxidation in fibroblasts were not found in patients with the muscular presentation. The presence or the absence of extramuscular symptoms could be correlated with the depth of CPT 2 defect. CPT 2 activity was indeed decreased by 90% in the patient with the hepato-cardio-muscular form, and by 75% in two previously studied patients with the muscular presentation of the disease. Variable exposure to environmental factors (prolonged exercise, fasting, infection, ...) could also account for the phenotypic heterogeneity of CPT 2 deficiency. This hypothesis is supported by the description of CPT 2 deficiency in siblings, one asymptomatic and the other with the "muscular" form of the disease (5).

III. HEPATIC CPT DEFICIENCY

The hepatic presentation of CPT deficiency was first described by Bougneres et al in 1981 (2). Since this date, nine other patients presenting with an acute liver failure have been shown to be affected with CPT deficiency (6, 8, 9, 10, 11, one unpublished patient from Dr. Cederbaum, and three unpublished patients from Drs. Haworth and Coates). Four patients' parents are first or second cousins. Onset of the disease always occurs below 18 months. The main clinical symptom is always a severe hypoketotic hypoglycemia triggered by prolonged fast and/or infection. During attacks, hepatomegaly with fatty liver is constant as well as elevated transaminases, prothrombin time, ammoniemia and triglyceridemia. Muscular or cardiac injury is not a feature of this disease but a transient myocardiopathy has been reported in one case. A renal tubular acidosis is noted in

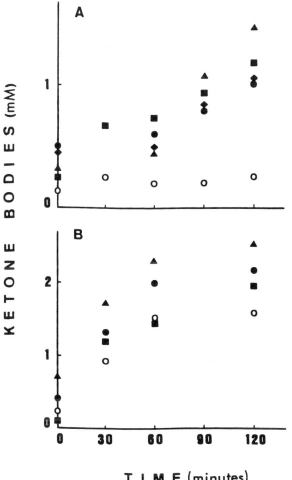

Figure 1

Time course of plasma ketone bodies levels after triglyceride loadings. Long-chain (A) and medium-chain (B) triglyceride loadings were performed in the patient (○) and controls (●, ■, ▲, ♦) as described in Methods. Ketone bodies are expressed as the sum of 3-hydroxybutyrate and acetoacetate.

two cases. Plasma total carnitine is normal or elevated with a normal free-esterified ratio. Urinary profile of organic acids is usually normal. Developmental assessment is normal in 7 cases and 3 patients are developmentally delayed. One patient suddenly died when 40-months old. The other patients are now aged from 3.5 to 13 years. A fasting test performed in 5 patients resulted in hypoglycemia with low blood ketones over a 15 hour fast (Figure 2). A medium-chain triglyceride loading test performed in 4 patients resulted in a normal rise of blood ketones (Figure 3). This pattern of blood ketone course focused on a liver defect of LCFA oxidation. This was confirmed in fibroblasts from all patients tested: palmitate oxidation was markedly decreased whereas octanoate was normally oxidized. In all patients' fibroblasts, CPT 1 activity was found deeply reduced and CPT 2 activity (as well as CPT 2 protein in the three patients tested) was normal (Table 2). That confirms that the hepatic presentation and the hepato-cardiomuscular presentation of CPT deficiency are

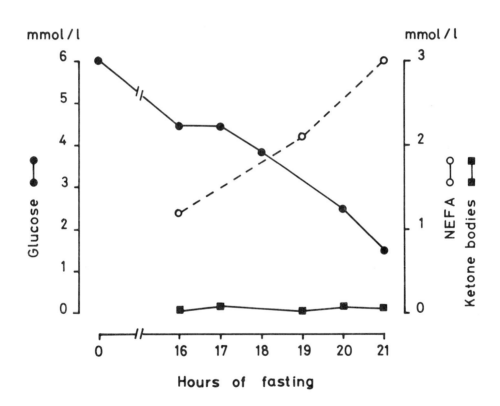

Figure 2

Plasma glucose, ketones, and free fatty acids during prolonged fast in a case of CPT 1 deficiency. Plasma glucose (○) declines but total ketone bodies (□) do not increase despite elevated non esterified fatty acids (▵) plasma concentrations.

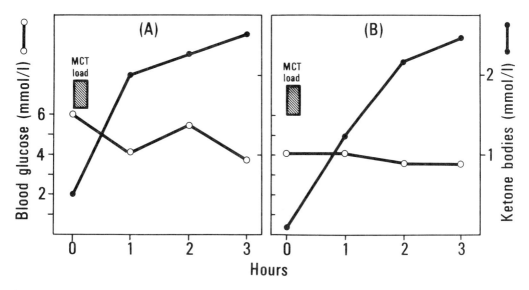

Figure 3

Effect of MCT load on plasma glucose (○) and ketone body concentrations (•) in a normal infant (A) and a patient with CPT 1 deficiency (B) in post-absorptive state.

distinct entities. The lack of muscular symptoms in the hepatic presentation was shown to result from a normal CPT 1 activity in two patients' skeletal muscle (12) (Table 3). This advocates distinct tissue-specific isoforms of CPT 1 in human liver and skeletal muscle, as suggested in the rat (13).

No information is available regarding cardiac CPT 1. Tissual specificity of CPT 1 makes unlikely a hepato-cardiomuscular presentation of CPT 1 deficiency. In this way, specific muscular and/or cardiac CPT 1 deficiency is to be considered.

TABLE 3
CPT Activities in Skeletal Muscle from CPT Deficient Patients

	Malonyl CoA 100 μM	Controls N = 10	"Muscular" n = 2	"Hepatic" 1 n = 3	"Hepatic" 2 n = 3
CPT I	−	19.6 ± 1.7	ND	25.0 ± 0.2	25.6 ± 4.2
	+	5.2 ± 0.8	ND	6.5 ± 0.4	6.1 ± .03
CPT II	−	9.5 ± 1.75	1.3; 1.4	9.9 ± 0.35	8.7 ± 1.0

REFERENCES

1. DI MAURO S, TREVISAN C (1982) Carnitine palmityltransferase (CPT) deficiency: a review. In SHOTLAND DL, ed, Disorders of the motor unit. John Wiley and Sons, New York. p. 657-666.

2. BOUGNERES PF, SAUDUBRAY JM, MARSAC C, BERNARD O, ODIEVRE M, GIRARD J (1981) Fasting hypoglycemia resulting from hepatric carnitine palmitoyl transferase deficiency. J Pediatr 98: 742-746.

3. DEMAUGRE F, BONNEFONT JP, COLONNA M, CEPANEC C, LEROUX JP, SAUDUBRAY JM (1991) Infantile form of carnitine palmitoyltransferase II deficiency with muscular and extramuscular symptoms. J Clin Invest 87: 859-864.

4. DI MAURO S, DI MAURO PMM (1973) Muscle carnitine palmitoyltransferase deficiency and myoglobinuria. Science 182: 929-931.

5. KELLY KJ, GARLAND JS, TANG TT, SHUG AL, CHUSID MJ (1989) Fatal rhabdomyoloysis following influenza infection in a girl with familial carnitine palmitoyltransferase deficiency. Pediatrics 84: 312-316.

6. DEMAUGRE F, BONNEFONT JP, MITCHELL G, NGUYEN-HOANG N, PELET A, RIMOLDI M, DI DONATO S, SAUDUBRAY JM (1988) Hepatic and muscular presentations of carnitine palmitoyltransferase deficiency: two distinct entities. Pediatr Res 24: 308-311.

7. DEMAUGRE F, BONNEFONT JP, CEPANEC C, SCHOLTE J, SAUDUBRAY JM, LEROUX JP (1990) Immunoquantitative analysis of human carnitine palmitoyltransferase I and I defects. Pediatr Res 27: 497-500.

8. BONNEFONT JP, HAAS R, WOLFF J, THUY LP, BUCHTA R, CARROLL JE, SAUDUBRAY JM, DEMAUGRE F, NYHAN WL (1989) Deficiency of carnitine palmitoyltransferase I. J Child Neurol 4: 198-203.

9. LAYWARD EM, HODGES S, SWIFT PGF, POLLITT RJ, BENNETT MJ, BARTLETT K (1987) Recurrent encephalopathy in a child of normal growth with a defect of long chain fatty acid oxidation. Abstract. SSIEM meeting, Sheffield.

10. MARANDIAN MH, SOLTANABADI A, RAKCHAN M, KOUCHANFAR A, FALLAH A (1987) Encéphalopathie augue et stéatose hépatique récurrentes avec activité normale de l'acyl-CoA deshydrogénase des acides gras à chaîne longue et moyenne. Arch Fr Pediatr 44: 369-371.

11. STANLEY CA, SUNARYO F, BONNEFONT JP, DEMAUGRE F, SAUDUBRAY JM (1991) Elevated carnitine levels in the hepatic form of carnitine palmityltransferase 1 deficiency. Subnitted to J Pediatr.

12. TIEN I, DEGAUGRE F, BONNEFONT JP, SAUDUBRAY JM (1989) Normal muscle CPT 1 and CPT 2 activities in hepatic presentation patients with CPT 1 deficiency in fibroblasts. Tissue specific isoforms of CPT 1? J Neurol Sci 92: 229-245.

13. WOELTJE KF, KUWAJIMA H, FOSTER DW, McGARRY JD (1987) Characterization of the mitochondrial carnitine palmitoyltransferase enzyme system. II. Use of detergents and antibodies. J Biol Chem 262: 9822-9827.

DISCUSSION OF THE PAPER

W. HAMILTON *(Danville, PA)*: My patient with long chain Fatty Acyl CoA Dehydrogenase defect, presented initially as possible new born sepsis, and a second time two months later. This patient had hypoglycemia, lack of ketosis, hepatomegaly and severe hypertrophic cardiomyopathy. The patient survived ICU and the cardiac muscle returned to near normal over 1-2 months, but recurrent episodes of hepato, cardiac, and occasional encephalopathy occurred with illnesses. These episodes, if severe, were associated recurrent hepatomegaly and hypertrophic cardiomyopathy. The patient is managed effectively now with vivonek feeding, and recurrent brief admissions for I.V. glucose therapy when vomiting occurs. He has mild muscle weakness and mixed CNS sequelae from an earlier cardiac arrest. He does not take L-carnitine.

P. SUPPLE *(Downers Grove, IL)*: Is there a possible dietary treatment of severe PCT II?

SAUDUBRAY: It is possible to try a low long chain fatty acid diet with a sufficient amount of TCM. However there is still no clinical evidence of the efficiency of this therapy. Another way could be to give snacks with uncooked cornstarch (2g/kg) by dose.

M. NOVAK *(Miami, FL)*: There are only a few mitochondria in lymphocytes and differences in the proportion in individual types of WBC. Can you explain the basics of your method (substrate conditions, etc.)?

SAUDUBRAY: We use a pellet of 10^6 fresh isolated lymphocytes and our basic experimental conditions are quite similar to those I describe in the fibroblast assay Pediat. Res 1982:16 877-881.

M. HALPERIN *(Canada M5B 1A6)*: In a patient with CPT II deficiency, given the large mass of muscle, the ATP lack, the plasma membrane that is leaky to large molecules (enzymes, myoglobin), why is acute hyperkalemia not a large problem?

SAUDUBRAY: It is actually a problem in a case of very severe access of myoglobinuria with renal insufficiency.

G. HUG *(Cincinnati, OH)*: Considering the markedly different clinical course of your CPT II deficient patient and ours reported in Pediatrics Res. 25:115A; 1989, is the generalized, severe CPT II deficiency with neonatal death versus that with infantile death one disease or two diseases?

SAUDUBRAY: I don't know - We need to investigate carefully at least each fibroblast cell line with the same methodology including kinetics analysis, western blot and pulse chase experiments. As soon as probes will be available it will be possible to look at the mutation. Extensive tissue studies would be welcomed each time it is possible. In your particular observation the high severity could be due to an almost absence of CPT II activity in most of the tested tissues and mainly in heart in which the defect has probably been responsible for the rapid death. In the patient we described in JCI 1991, the residual activity in fibroblasts was about 10%. Unfortunately, we had no opportunity to study other tissue, (heart, muscle, and liver).

ABSTRACTS TO PART III

EVIDENCE FOR THE EXISTENCE OF AN IDENTICAL CARNITINE PALMITOYLTRANSFERASE ENZYME IN RAT HEART AND LIVER. G.P. Heathers, M.Zs. Kozak, R. Daniel, K-S. Huang, S. Li, J.E. Smart, A.J. Higgins, and W. Levin. Departments of Cardiovascular Research and Protein Biochemistry, Hoffmann-La Roche Inc., Nutley, NJ 07110.

Conflicting evidence has been reported describing the existence of one or more carnitine palmitoyltransferase (CPT) enzymes in heart, skeletal and liver tissue. We report evidence indicating that an identical CPT enzyme exists in both heart and liver tissues. These studies made use of an anti-CPT polyclonal antibody (prepared from rat liver enzyme and kindly provided by Dr. Paul Brady, University of Minnesota) and consisted of three types of experiments. First, mitochondrial fractions from rat heart and liver were subjected to SDS-PAGE and the resolved proteins transferred to nitrocellulose. In both heart and liver an identical protein of $M_r = 68000$ was detected with the anti-CPT antibody and goat anti-rabbit $F(ab)_2$ conjugated to HRP. Second, SDS-PAGE of heart and liver mitochondrial extracts was run as above and gel corresponding to the $M_r = 68000$ region partially digested with CNBr. Digested samples were run on SDS-PAGE, transferred to nitrocellulose, probed using the anti-CPT polyclonal antibody and identified with anti-rabbit $F(ab)_2$ conjugated with AP. An identical peptide pattern for both heart and liver enzymes was demonstrated. Finally, solubilized (Triton X100, 0.5%) mitochondrial preparations from heart and liver (containing all measurable CPT activity) were immunodepleted by incubation with anti-CPT antibody and protein G-conjugated beads. An identical depletion of CPT activity from the supernatant was seen for both heart and liver preparations. These experiments indicate that an identical CPT enzyme exists in both heart and liver tissue. However, it is not known whether additional CPT enzymes are present in either or both tissues.

KINETIC AND INHIBITORY MECHANISMS OF CARNITINE ACYLTRANSFERASE IN INTACT MITOCHONDRIA FROM RAT HEART. Anthony T. Greway and Marygrace E. McDonald. Department of Pharmacology, Cardiovascular Research, Hoffmann-LaRoche, Inc., 340 Kingsland St., Nutley, NJ 07110.

The kinetic mechanism of carnitine acyltransferase, as expressed on the outer surface of the inner mitochondrial membrane in intact mitochondria from rat heart, CAT-I or CPT_{overt}, was investigated through the use of dead-end inhibition studies with both substrate analogs (des-hydroxy-carnitine and malonylCo-A) and products (palmitoylcarnitine and Co-A). The data gathered was consistent with the enzyme following an ordered Bi-Bi kinetic mechanism with palmitoylCo-A (palCo-A) binding prior to carnitine and palmitoylcarnitine leaving the active site before Co-A. The kinetic constants determined for the enzyme performing catalysis in the forward direction are: $K_{carn} = 270 \pm 20$ μM, $K_{palCo-A} = 25 \pm 10$ μM and $V_{max} = 25 \pm 8$ (pmol/min)/μg. Scatchard binding studies with $[2-^{14}C]$malonylCo-A indicate that malonylCo-A binds to the intact mitochondria at two distinct sites, a high affinity site ($K_{D1} = 0.1$ μM, $N_1 = 40$ pmol/mg), and a low affinity site ($K_{D2} = 100$ μM, $N_2 = 1000$ pmol/mg). PalCo-A was found to be competitive with $[2-^{14}C]$malonylCo-A at only the low affinity site. In the inhibition studies with malonylCo-A, the maximum inhibition was found to decrease with time (over 7 days); decreasing more rapidly for that due to binding at the high affinity site (55% to 10%), than that due to binding at the low affinity site (60% to 43%). During the time course studied there was no loss in binding of malonylCo-A to the mitochondria and no decrease in the maximum inhibition from des-hydroxy-carnitine. This indicates that the loss of inhibition by malonylCo-A at saturation of the high affinity site may be due to the dissociation of a regulatory subunit (or a high affinity malonylCo-A binding protein) from the catalytic polypeptide, which does not result in alterations in the catalytic properties of the enzyme.

STIMULATION OF CARNITINE PALMITOYLTRANSFERASE (CPT) ACTIVITIES OF DIFFERENT SUBCELLULAR FRACTION BY ACYL-CoA BINDING PROTEIN (ACBP) <u>A.K.M. Jalaluddin Bhuiyan and Shri V. Pande</u>, Clinical Research Institute of Montreal, Montral, Quebec, Canada H2W 1R7

Work related to the malonyl-CoA inhibitable CPT has implicated malonyl-CoA inhibition in the regulation of the delivery of acyl-CoA to the β-oxidation. With palmitoyl-CoA, this inhibition <u>in vitro</u> requires the presence of BSA as an acyl-CoA binding agent. We have reported (see FASEB J. Abstract, 1991) that BSA markedly stimulated the CPT activity with octanoyl-CoA and increased the malonyl-CoA sensitivity of CPT_0 of the outer mitochondrial membrane (OMV). This indicated that the protein bound acyl-CoA may be the preferred substrate for the above expression. Thus, we examined the role of purified homogeneous ACBP (10 kDa protein) in rat liver CPT system. Our data support the notion that ACBP could serve as the physiological donor of the acyl-CoA to the CPT system. Thus, (a) when a low near physiological concentration of palmitoyl-CoA was offered to the CPT of intact mitochondria, OMV and peroxisomes, in the absence of BSA, inclusion of ACBP markedly increased the malonyl-CoA inhibition. (b) When the palmitoyl-CoA concentration was high, the basal CPT activities of the above subcellular fractions, of the inner mitochondrial membrane, and the microsomes were increased several fold by ACBP. We are now examining whether fatty acid binding protein may also have a role in the CPT system. (Supported by grants from the Medical Research Council of Canada and the Quebec Heart Foundation).

FATAL DEFICIENCY OF CARNITINE PALMITOYL TRANSFERASE II (CPT II) IN LIVER, HEART AND MUSCLE OF A NEWBORN GIRL. <u>G. Hug, M. Tsoras, M. Ryan, K. Bove and S. Soukup.</u> The Children's Hospital Medical Center, Cincinnati, OH 45229.

A girl with deficient CPT II died at age 5 days of encephalo-cardiomyopathy. In her heart, liver and muscle, fat was excessive, activities of palmitoyl CoA synthase, carnitine acetyl transferase, 4 acyl CoA dehydrogenases, CPT I were normal. The activity of CPT II was reduced to < 6% (liver, muscle), or was undetectable (heart); carnitine was low, acylcarnitine high. She was well for 2 d, then developed lethargy, hypoglycemia, seizures, hepatomegaly, hypotonia and died 3 d later in cardiac arrest (Hug, G. et al. Pediatr Res 25:115A;1989). Other investigators observed a boy, healthy until 3 mo when he had a metabolic crisis. Thereafter he was well until he died suddenly at age 17 months. CPT II was reduced to < 10% in his cultured fibroblasts. CPT in other tissue was not assayed, but it was assumed to be < 10% and to result in the boy's clinical picture (Demaugre, F. et al. J Clin Invest 87:859;1991). However, the different clinical course of the two patients may reflect differences in CPT tissue impairment, and an understanding of different phenotypes in CPT deficient patients requires parenchymal tissue analyses.

PRESENCE OF CARNITINE TRANSPORTER IN THE COMPLEX OF CARNITINE PALMITOYLTRANSFERASE—β-OXIDATION ENZYMES ELUTED FROM ANTI-CPT ANTIBODY—SEPHAROSE COLUMN. M.S.R. Murthy and L.L. Bieber, Dept. of Biochemistry, Michigan State University, East Lansing, MI 48824.

Malonyl-CoA sensitive and insensitive carnitine palmitoyltransferase (CPT) activity of heart mitochondria, in combination with three of the four enzymes of β-oxidation machinery, can be bound to and eluted from anti-CPT antibody-Sepharose column (Kerner and Bieber, *Biochemistry* **29**, 4326, 1990). We now report that this complex of proteins containing CPT and β-oxidation enzymes also contains carnitine transporter activity. The transporter activity was measured after reconstituting the solubilized mitochondrial protein and the eluted protein complex from anti-CPT antibody column into asolectin liposomes by freeze-thaw sonication. There was an ~ 2-fold increase in the specific activity of the transporter in the CPT—β-oxidation complex as compared to the starting extract. The presence of the carnitine transporter in this complex was further confirmed by using anti-carnitine transporter antibodies in the immunoblot experiments. These anti-transporter antibodies recognized a 32 kDa protein, corresponding to the purified carnitine transporter, and also a 68 kDa protein, presumably CPT, in the immunoblots of proteins eluted from CPT-antibody column. These results suggest that CPT and the carnitine transporter may be associated with each other in the inner membrane of mitochondria. Such association may facilitate the channeling of acyl esters of carnitine between these two proteins. (Supported in part by NIH grant DK18427)

PART IV

CARNITINE AND CARDIOLOGY

Introduction to Carnitine and the Heart

Ann Willbrand, Ph.D.
Assistant Professor of Chemistry
Chair, Department of Physical Science
University of South Carolina at Aiken

Second only to the male reproductive tract, the heart has the highest levels of carnitine. The two known functions of carnitine, carrying long chain fatty acids into the mitochondria and generating free coenzyme A (CoA) from acyl-CoA, are of critical importance in normal heart metabolism. Early studies indicated that patients with diphtheric cardiomyopathy had reduced carnitine levels due to a defect in the transport of carnitine into the heart (1). Later a series of patients with low plasma carnitine and a heart condition known as fibroelastosis were identified (2). Treatment of these patients with long term oral carnitine therapy resulted in dramatic improvement. The necessity of carnitine administration over several years in these patients has been demonstrated (3).

From the early studies many questions have arisen about the role of carnitine in treatment/prevention of various pathological conditions of the heart ranging from acute ischaemia to congestive heart failure. Carnitine supplementation has been found to be beneficial in the treatment of many cardiac conditions. In the following chapters, current investigations into carnitine and its role in the heart are discussed. The first chapter contains a detailed discussion of carnitine studies with human subjects, while the following two chapters describe basic research with animal models.

The last chapter, written by the editor, addresses the use of carnitine to prevent the cardiotoxic side effects of adriamycin (doxorubicin), a widely used chemotherapeutic agent. Adriamycin is highly effective against many different types of malignancies, but unfortunately many patients develop cardiomyopathy, and sometimes cardiac failure. Carnitine appears to be able to prevent these cardiac problems associated with large doses of adriamycin.

1.   Bohmer, T. and Molstad, P. In Frenkel, R.A. and McGarry, J.D. (eds) Carnitine Biosynthesis, Metabolism, and Functions, Academic Press, NY Pp 73-90 (1980).
2.   Tripp, M.E., Katcher, M.L., Peters, H.A., Gilbert, E.F., Arya, S., Hodach, R.J. and Shug, A.L. New England Journal of Medicine 305:385-390 (1981).
3.   Tripp, M.E. and Shug, A.L. Pediatric Research 19:A135 (1985).

CARNITINE AND MYOCARDIAL FUNCTION

Mary Ella M. Pierpont M.D. Ph.D.
University of Minnesota
Minneapolis, Minnesota

I.   INTRODUCTION

L-carnitine has been known for many years to be the essential cofactor for mitochondrial transport of fatty acid groups. For this reason, investigators have been very interested in the importance of regulation of fatty acid metabolism and the function of the myocardium.[1]. Sites at which regulation of fatty acid metabolism has been and continues to be investigated include: 1. Fatty acid uptake and activation; 2. Mitochondrial transport of fatty acyl residues; 3, Beta oxidation; 4. Carnitine availability and transport; 5. Availability of oxygen; and 6. Increased myocardial work. In this chapter, aspects of myocardial carnitine availability and transport are addressed in clinical states and animal models of congestive heart failure.

Very little is known about the transport of carnitine into the human heart. In animals, the cardiac carnitine concentration is as much as 40-fold higher than in plasma, thus requiring an active transport process.[2] Under conditions of physiological serum carnitine levels, 80% of the uptake of total myocardial carnitine in isolated rat hearts is related to a carrier-mediated transport system which depends on the extracellular sodium concentration.[3] A carnitine binding protein has been isolated which is associated with the plasma membrane of rat cardiac tissue and has properties which suggest its involvement in carnitine transport.[4]

Abnormalities of carnitine transport into the heart resulting in myocardial carnitine deficiency have been described in hamsters with hypertrophic cardiomyopathy.[5,6] and dilated cardiomyopathy.[6,7] These animals have low myocardial carnitine, but normal or elevated plasma carnitine is present along with elevated liver carnitine, suggesting that these animals have normal synthetic capacity, but altered means of transporting carnitine into myocardial cells. The hamster myocardial deficiency of carnitine can be improved significantly by L-carnitine treatment of those hamsters with dilated cardiomyopathy, but not the hypertrophic cardiomyopathy.[7]

Mechanical performance of the hearts of both hamster cardiomyopathies was improved by L-carnitine supplementation.

In humans, deficiency of carnitine in plasma and tissues has been reported for many years, and a carnitine transport abnormality has been suspected. In a number of children, previously called "systemic carnitine deficiency" or "primary carnitine deficiency", there is a severe cardiomyopathy associated with very low plasma carnitine and extremely low levels of skeletal muscle carnitine (Table 1). These children were thought, but not proven, to have a carnitine transport defect, including a loss of carnitine in the urine due to deficient renal tubular transport or reabsorption.[11] In 4 of the 5 children in Table 1, L-carnitine supplementation was lifesaving, with one child having died prior to the diagnosis of carnitine deficiency.

TABLE 1

Cases of Carnitine Deficiency and Cardiomyopathy

Author	Age (Mo.)	Sex	Presenting Feature	Plasma Total Carnitine	% Normal Muscle Carnitine
Chapoy et al (8)	3	M	Coma	4.8	0.5
Tripp et al (9)	11	F	Cardiac Failure	4.8	1.0
Tripp et al (9)	26	M	Cardiac Failure	4.5	1.0
Waber et al (10)	40	M	Cardiac Failure	4.2	2.0
Rodrigues-Pereira (11) et al	18	M	Cardiac Failure	1.8	1.5

In 1988, Treem et al reported evidence that a child with coma and thickening of the left ventricular walls had markedly decreased ability to transport carnitine into cultured fibroblasts, providing clear evidence for an abnormal carnitine transport mechanism.[12] This child had extremely low plasma and skeletal muscle carnitine (Table 2). Clinical improvement has ensued on L-carnitine therapy although the skeletal muscle carnitine level has increased from 0.5% to only 3% of normal. Others (Ericksson et al [15]; Tein et al [14]) have demonstrated concentration dependent differences in carnitine uptake in similar patients with cardiomyopathy and carnitine deficiency. In some of these children, a renal tubular carnitine reabsorption defect has also been demonstrated.[13,14] All have shown significant improvement from L-carnitine supplementation.

TABLE 2

Cases of Carnitine Transport Defect

Cases	Age (mo)	Sex	Presenting Feature	Total Plasma Carnitine nmol/ml	% Normal Muscle Total Carnitine
Treem et al (12)	3.5	F	Coma	0-2.2	0.5
Eriksson et al (13)	48	F	Cardiac Failure	<3%	1.0
Tein et al (14)	1	F	Cardiac Failure	19	4.7
Tein et al (14)	17	M	Failure to Thrive	1.2	5.5
Tein et al (14)	30	F	Coma	0	N.D.
Tein et al (14)	2	F	Failure to Thrive	9	4.7
Present Case 1	78	M	Cardiac Failure	1.0	0.02
Present Case 2	66	F	No Symptoms	1.2	0.01

II.   RESTORATION OF NORMAL MYOCARDIAL FUNCTION IN MEMBRANE CARNITINE DEFECT

A.   CASE DESCRIPTIONS

In Table 2 are depicted two new cases of carnitine transport defect described here. Case 1 is a 6 1/2 year old boy who developed congestive heart failure over a period of 6 weeks. He had been previously healthy except for asthma. His physical examination revealed tachypnea and tachycardia. There was a grade 2/6 long systolic murmur at the cardiac apex representing mitral regurgitation. The liver was enlarged to 5 cm below the right costal margin. Mild hypotonia and proximal muscle weakness were evident. A chest x-ray revealed markedly increased cardiac size and evidence of pulmonary edema. An electrocardiogram revealed left ventricular hypertrophy with peaked T waves in the left precordial leads. Echocardiography revealed severely reduced cardiac contractility and enlargement of the left ventricular internal dimension. The left ventricular ejection fraction was <20%. Plasma and skeletal muscle total carnitine levels were 2% and 0.02% of normal respectively (Table 2). Lipid storage was widespread in skeletal muscle (Figure 1).

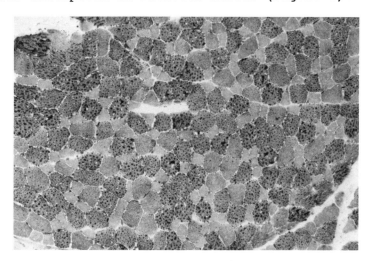

Figure 1:   Photomicrograph of skeletal muscle of case 1.
          Oil red 0 stain.

Case 2 is the asymptomatic sister of case 1 who was found to
have a cardiac murmur. She had previously been entirely healthy.
There was a grade 2/6 holosystolic murmur heard best at the
cardiac apex. A chest x-ray (Figure 2A) revealed moderately
severe cardiomegaly. An electrocardiogram showed left
ventricular hypertrophy with peaked T waves similar to her
brother. Echocardiography revealed a dilated cardiomyopathy with
reduced ejection fraction and increased left ventricular internal
dimension. Plasma and skeletal muscle total carnitine levels
were severely reduced, similar to case 1 (Table 2).

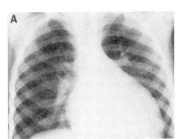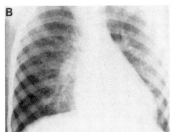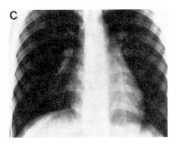

Figure 2: Chest x-rays, Case 2
      A. At diagnosis.
      B. After 1 month L-carnitine therapy.
      C. One year of L-carnitine therapy.

Skin fibroblasts from both children were obtained for
studies of carnitine transport. These studies revealed that Case
1 and Case 2 each had very low transport velocity at 5μ mol/L
L-carnitine (0.016 and 0.019 compared to control values of
0.80-120 pmol/min/mg protein). Fibroblasts of both the mother
and the father had intermediate transport velocities at 5 μmol/L
L-carnitine (0.49 and 0.51 pmol/min/mg protein) suggesting
heterozygosity for a recessive gene.

B.   MYOCARDIAL IMPROVEMENT WITH L-CARNITINE
Both Case 1 and Case 2 were begun on 100 mg/kg/day of
L-carnitine supplementation. After 1 month, symptoms of
congestive heart failure had subsided in Case 1. Both children,
by one month of L-carnitine supplementation, had an obvious
decrease in cardiac silhouette (Figure 2B) and by 1 year of
supplementation, the cardiac size was normal (Figure 2C). Plasma
total carnitine increased to a normal range in both children by 1
month of supplementation (Figure 3). Doses up to 150 mg/kg/day
of L-carnitine have been necessary to maintain normal plasma

carnitine, an observation also documented in other children with carnitine membrane transport defects.[14]

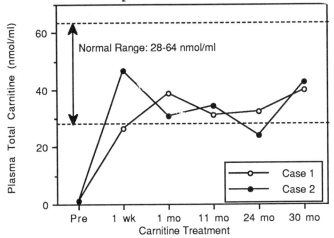

Figure 3: Plasma total carnitine during L-carnitine therapy.

   Echocardiographic measurements of cardiac function in these 2 children have demonstrated sustained improvement in myocardial function over 2 years of L-carnitine therapy. The percent of fractional left ventricular shortening (Figure 4), a measurement of ventricular contractility, improved from extremely low to the normal range in Case 2 by 1 week of supplementation. Case 1, who had significant symptoms of congestive heart failure, took longer to improve, but the percent of fractional left ventricular shortening increased to normal after 2 months of supplementation and has revained normal for over 2 years. In both children, the left ventricular end-diastolic dimension was much greater than normal prior to L-carnitine supplementation (Figure 5) and this dimension slowly and progressively decreased to normal over the 2 years of supplementation.

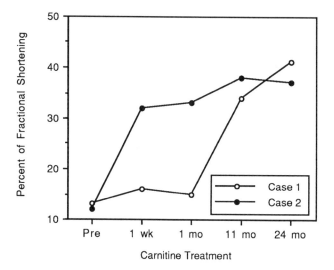

Figure 4:   Improvement in fractional left ventricular shortening with L-carnitine therapy.

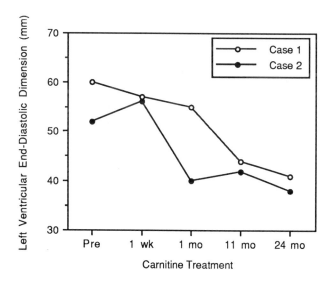

Figure 5:   Decrease in left ventricular end-diastolic dimension with L-carnitine therapy.

These data suggest that the dramatic improvement in myocardial function in both of these children is directly related to their L-carnitine supplementation. Ordinarily, children with the degree of cardiomyopathy that was present in these two children would have a 50% mortality over the ensuing year. The impressive manner of their improvement suggests L-carnitine was life-saving. We do not know if these children also have impairment of carnitine transport into the heart tissue or simply into fibroblasts and muscle with a renal tubular reabsorption defect. Repeat muscle biopsy and cardiac biopsy (serial) with measurement of carnitine levels would help to answer these questions.

III. MYOCARDIAL CARNITINE IN END-STAGE CONGESTIVE HEART FAILURE

Because of the findings of cardiomyopathy associated with carnitine deficiency,[9-11] a study was initiated to evaluate whether individuals with end-stage heart failure might also have decreased carnitine in their myocardium.[16] In some animal models of congestive heart failure, myocardial carnitine deficiency has been documented. These include Syrian hamsters,[5-7,17] guinea pigs with diphtheria,[18] rats with pressure-overload myocardial hypertrophy[19] and dogs with regional myocardial ischemia caused by surgical ligations.[20] Such findings in animals suggested that myocardial carnitine deficiency could be found in humans with congestive heart failure of other causes than cardiomyopathy.

60 patients were studied who were undergoing orthotopic cardiac transplantation. After surgical removal of the recipient heart, a transmural section of the left ventricular free wall was obtained for carnitine measurement. Similar left ventricular samples were obtained at autopsy from 5 control hearts (no malformations or evidence of coronary artery disease) who died of noncardiac causes. Five biopsy samples of left ventricular myocardium were obtained from patients with normal cardiac ejection fractions undergoing coronary artery bypass. Blood plasma samples were obtained from 32 of the 60 subjects within 24 hours prior to cardiac transplantation. Plasma and myocardial carnitine were measured by a radioisotopic method as previously described.[21,22]

The patients ranged in age from 6 months to 64 years (mean 38± 17 years) at the time of transplantation (Table 3). All patients undergoing cardiac transplantation had significantly reduced left ventricular ejection fractions and increased pulmonary arterial wedge pressures (Table 3). Of those undergoing cardiac transplantation, 35 had dilated cardiomyopathy, 14 had coronary artery disease with poor cardiac function, 8 had myocarditis and 3 had other forms of heart failure (2 with rheumatic heart disease and one with Chagas disease). Patients with coronary artery disease were significantly older than those with dilated cardiomyopathy (p <0.001) or myocarditis (p<0.001).

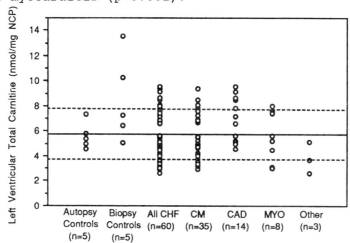

Figure 6: Left ventricular total carnitine measurements in end-stage heart failure. CHF=heart failure patients; CM= cardiomyopathy; CAD= coronary artery disease; MYO= myocarditis.

Figure 6 shows the striking findings of marked variability in the left ventricular total carnitine levels in the different groups of end-stage heart failure patients. The mean values of total carnitine for the control autopsy group (5.7 ± 1.0 nmol/mg NCP) and the transplant group (5.8± 1.8 nmol/mg NCP) were not significantly different. However, of the entire group of 60 hearts of patients with heart failure, 10 (17%) had left ventricular total carnitine levels that were more than 2 standard deviations below the mean of the 5 autopsy hearts. Of the 10 hearts with very low myocardial carnitine, 6 had dilated cardiomyopathy, 2 had acute myocarditis, 1 had rheumatic heart disease, and 1 had Chagas disease. No patient with coronary artery disease had low myocardial carnitine. There was a wide spread of values for the 5 left ventricular samples of the biopsy

TABLE 3

Hemodynamic Measurements in End-Stage Heart Failure

Heart Transplant Patient Diagnosis	N	Age (yrs)	Ejection Fraction (%)	Pulmonary Arterial Wedge (mmHg)	Cardiac Index (L/min/m²)	Pulmonary Arteriolar Resistance (dynes sec cm⁻⁵)
Biopsy Controls	5	65 ± 8	59 ± 11			
All Congestive Heart Failure	60	38 ± 17	18 ± 8	27 ± 7	2.0 ± 0.5	233 ± 125
Cardiomyopathy	35	35 ± 17	19 ± 10	27 ± 7	2.1 ± 0.5	212 ± 95
Coronary Artery Disease	14	51 ± 8	16 ± 6	27 ± 8	1.9 ± 0.4	270 ± 140
Myocarcitis	8	25 ± 17	18 ± 7	28 ± 9	2.2 ± 0.8	251 ± 207
Other	3	48 ± 10	17 ± 5	25 ± 3	1.7 ± 0.7	230 ± 96

group, with the mean cardiac total carnitine higher in the biopsy group compared to the autopsy group. These values suggest that there may be some loss of myocardial carnitine after death and prior to autopsy. The left ventricular biopsy group, on the other hand, does not represent completely normal hearts, since these patients were undergoing coronary artery bypass for coronary artery disease. They did, however, have normal left ventricular ejection fractions which were much greater than those undergoing transplantation.

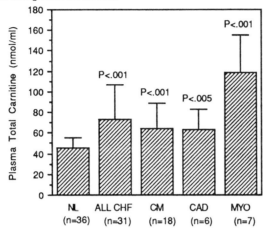

Figure 7: Plasma total carnitine in end-stage heart failure. Abbreviations as in Figure 6.

None of the 32 patients in whom blood samples were obtained prior to surgery had plasma free or total carnitine deficiency. This is in contrast to other series where deficiency of plasma carnitine was detected in 6 of 27 adults with dilated cardiomyopathy in one study and 2 of 25 in another.[23,24] In the present study, mean plasma total carnitine levels were significantly elevated in all different types of heart failure tested in our study (Figure 7). Similar elevated plasma carnitine values have also been described by others in humans with congestive heart failure of various causes,[23-26] and in animals with cardiomyopathy or heart failure including turkeys and hamsters.[21,7]

The etiology of the plasma carnitine elevation in end-stage heart failure is not known precisely. It has previously been suggested that decreased renal excretion of carnitine may be related to the elevated plasma levels,[23] but no correlation between elevated plasma carnitine and creatinine clearance was found in 20 of the patients in whom both were measured.[16] It would appear that decreased renal function alone cannot completely explain the elevated plasma and other mechanisms such as decreased tissue transport or loss of carnitine from tissues must be considered.

IV. MYOCARDIAL CARNITINE DEPLETION IN CONGESTIVE
     HEART FAILURE INDUCED BY INCESSANT TACHYCARDIA

For many years it has been observed in humans that rapid incessant tachycardias can produce severe congestive heart failure. If the arrhythmia is interrupted and the individual converted to a normal cardiac rhythm and rate, the congestive heart failure resolves.[27] We hypothesized, based on our findings in humans with end-stage congestive heart failure undergoing cardiac transplantation, that decreased myocardial carnitine and elevated plasma carnitine would be detected in an animal model of incessant tachycardia.

In the past few years considerable interest has been generated by animal models of chronic rapid cardiac pacing which produce congestive heart failure [28,29]. New information on neurohumeral,[30,31] metabolic,[32] and cardiovascular reflex[33] responses to heart failure have been obtained. The model chosen for our investigation was the dog with persistent rapid ventricular pacing.[34]

Twelve mongrel dogs were studied.[34] Venous blood was obtained at baseline. Hemodynamic measurements (mean arterial pressure, cardiac output and pulmonary arterial wedge pressure) were obtained at the time of surgical implantation of a right ventricular transvenous pacemaker. Venous blood was obtained at 7,14 and 19 days of rapid ventricular pacing at a rate of 250 beats/minute. At 19 days, repeat hemodynamic measurements were obtained prior to sacrifice. Myocardial tissue and venous blood were analyzed for carnitine measurements.[21,22]

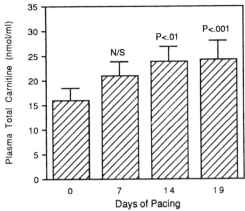

Figure 8: Plasma total carnitine measurements in dogs (n=12) with rapid-pacing induced heart failure.

Figure 8 reveals the serial analysis of plasma total carnitine illustrating that it progressively rises over 19 days of ventricular pacing. At 19 days, the dogs had significant congestive heart failure as evidenced by low mean arterial pressure, reduced cardiac output and elevated pulmonary arterial wedge pressures compared to baseline[34]. Also at 19 days, the plasma total carnitine was elevated ($p < 0.001$) compared to baseline.[34] Plasma free carnitine was similarly increased ($p < 0.001$) (data not shown).

With respect to myocardial measurements, a control group of 8 normal dogs was used for comparison. Myocardial total carnitine, free carnitine, short chain acylcarnitine and long chain acylcarnitine were all significantly lower in the left ventricle of paced dogs compared to control dogs (5.6 $\pm$ 1.5 vs 14.1 $\pm$ 3.5 nmol/mg NCP for total carnitine, $p < 0.001$).[34]

These results document that, associated with congestive heart failure in the dog, there is an abnormal plasma to myocardial carnitine ratio. There are significantly less total and free myocardial carnitine than normal, accompanied by elevation of the plasma total and free carnitine. This suggests either interference with the normal myocardial carnitine transport mechanism or loss of carnitine from the myocardial cells. The low levels of myocardial carnitine may also cause an unfavorable metabolic situation for the failing heart, making it less able to utilize a preferred substrate (fatty acids) and thereby further decreasing the cardiac efficiency.

ACKNOWLEDGEMENTS

The author would like to thank the other participants of the above described studies: D. Judd, G. Breningstall, A. Singh, I.F. Goldenberg, M.T. Olivari, W.S. Ring, G.L. Pierpont, J.E. Foker, C.A. Stanley, and S. Ewald.

REFERENCES

1. Morgan, H.E., and Neely, J.R., Metabolic regulation and myocardial function, in The Heart, Arteries and Veins, Vol.1, Hurst, W.J., Ed., McGraw Hill, New York, 1986, 85.
2. Siliprandi, N., Ciman, M., and Sartorelli, L., Myocardial carnitine transport, in Lipid Metabolism in the Normoxic and Ischaemic Heart, Stam, H., and vanderVusse, G.J., Eds., Steinkopff Verlag, Darmstadt, 1978, 53.
3. Vary, T.C., and Neely, J.R., Sodium dependence of carnitine transport in isolated perfused adult rat hearts, Am. J. Physiol., 244, H247, 1983.
4. Cantrell, C.R., and Borum, P.R., Identification of a cardiac carnitine binding protein, J. Biol. Chem., 257, 10599, 1982.
5. York, C.M., Cantrell, C.R., and Borum, P.R., Cardiac carnitine deficiency and altered carnitine transport in cardiomyopathic hamsters, Arch. Biochem. Biophys., 221, 526, 1983.
6. Whitmer, J.T., Energy metabolism and mechanical function in perfused hearts of Syrian hamsters with dilated or hypertrophic cardiomyopathy, J. Mol. Cell. Cardiol., 18, 307, 1986.
7. Whitmer, J.T., L-carnitine treatment improves cardiac performance and restores high-energy phosphate pools in cardiomyopathic Syrian hamsters, Circ. Res., 61, 396, 1987.
8. Chapoy, P.R., Angelini, C., Brown, W.J., Stiff, J.E., Shug, A.L., and Cederbaum, S.D., Systemic carnitine deficiency – a treatable inherited lipid storage disease presenting as Reye's Syndrome, N. Eng. J. Med., 303, 1389, 1980.
9. Tripp, M.E., Katcher, M.L., Peters, H.A., Gilbert, E.F., Arya, S., Hodach, R.J., and Shug, A.L., Systemic carnitine deficiency presenting as familial endocardial fibroelastosis: A treatable cardiomyopathy, N. Eng. J. Med., 305, 385, 1981.
10. Waber, L.J., Valle, D., Neill, C., DiMauro, S., and Shug, A., Carnitine deficiency presenting as familial cardiomyopathy, a treatable defect in carnitine transport, J. Pediatr., 101, 700, 1982.
11. Rodrigues-Pereira, R., Scholte, H.R., Luyt-Houwen, I.E.M., and Vaandrager-Verduin, M.H.M., Cardiomyopathy associated with carnitine loss in kidney and small intestine, Eur. J. Pediatr., 148, 193, 1988.
12. Treem, W.R., Stanley, C.A., Finegold, D.A., Hale, D.E., and

Coates, P.M., Primary carnitine deficiency due to a failure of carnitine transport in kidney muscle and fibroblasts. <u>N. Engl. J. Med.</u>, 319, 1331, 1988.

13. Eriksson, B.O., Lindstedt, S, and Nordin, I., Hereditary defect in carnitine membrane transport is expressed in skin fibroblasts, <u>Eur. J. Pediatr.</u>, 147, 662, 1988.

14. Tein, I., DeVivo, D.C., Bierman, F., Pulver, P., DeMeirleir, L.J., Cvitanovic-Sojat, L., Pagon, R.A., Bertini, E., Dionisi-Vici, C., Servidei, S., and DiMauro, S., Impaired skin fibroblast carnitine uptake in primary systemic carnitine deficiency manifested by childhood carnitine-responsive cardiomyopathy, <u>Pediatr. Res.</u>, 28, 247, 1990.

15. Eriksson, B.O., Gustafson, B., Lindstedt, S., and Nordin, I., Transport of carnitine into cells in hereditary carnitine deficiency, <u>J. Inher. Metab. Dis.</u>, 12, 108, 1989.

16. Pierpont, M.E.M., Judd, D., Goldenberg, I.F., Ring, W.S., Olivari, M.T., and Pierpont, G.L., Myocardial carnitine in end-stage congestive heart failure, <u>Am. J. Cardiol.</u>, 64, 56, 1989.

17. Yamashita, T., Hayashi, H., Kaneko, M., Kamikawa, T., Kobayashi, A., Yamazaki, N., Miura, K., Shirasawa, H., and Nishimura, M., Carnitine derivatives in hereditary cardiomyopathic animals, <u>Jpn. Heart J.</u>, 26, 833, 1985.

18. Bressler, R., and Wittels, B., The effect of diphtheria toxin on carnitine metabolism in the heart, <u>Biochim. Biophys. Acta,</u> 104, 39, 1965.

19. Reibel, D.K., Uboh, C.E., and Kent, R.L., Altered coenzyme A and carnitine metabolism in pressure-overload hypertrophied hearts. <u>Am. J. Physiol.</u>, 244, H839, 1983.

20. Shug, A.L., Thomsen, J.H., Folts, J.D., Bittar, N., Klein, M.I., Koke, J.R., and Huth, P.J., Changes in tissue levels of carnitine and other metabolites during myocardial ischemia and anoxia, <u>Arch. Biochem. Biophys.,</u> 187, 25, 1978.

21. Pierpont, M.E.M., Judd, D., Borgwardt, B., Noren, G.R., Staley, N.A., and Einzig, S.E., Carnitine alterations in spontaneous and drug-induced turkey congestive cardiomyopathy, <u>Pediatr. Res.</u>, 19, 415, 1985.

22. Dunnigan, A., Pierpont, M.E., Smith, S.A., Breningstall, G., Benditt, D.G., and Benson, D.W., Cardiac and skeletal myopathy associated with cardiac dysrhythmias, <u>Am. J. Cardiol.</u>, 53, 731, 1984.

23. Feldman, A.M., Waber, L.J., DeMent, S.H., Olson, J.L., and Baughman, K.L., Plasma carnitine levels in adults with dilated cardiomyopathy, <u>Heart Failure</u>, 3, 39, 1987.

24. Tripp, M.E., and Shug, A.L., Plasma carnitine concentrations in cardiomyopathy patients, <u>Biochem. Med.</u>, 32, 199, 1984.

25. Conte, A., Hess, O.M., Maire, R, Gautschi, K., Brogli, S., Knaus, U., and Krayenbuhl, H.P., Klinische Bedeutung des

Serumcarnitins fur den Verlauf und die Prognose der dilatativen Kardimyopathie, <u>Z. Kardiol.</u>, 76, 15, 1987.

26. Regitz, V., Shug, A.L., and Fleck, E., Defective myocardial carnitine metabolism in congestive heart failure secondary to dilated cardiomyopathy and to coronary, hypertensive and valvular heart diseases, <u>Am. J. Cardiol.</u>, 65, 755, 1990.

27. Packer, D.L., Bardy, G.H., Worley, S.J., Smith, M.S., Cobb, F.R., Coleman, R.E., Gallagher, J.J., and German, L.D., Tachycardia-induced cardiomyopathy: A reversible form of left ventricular dysfunction, <u>Am. J. Cardiol.</u>, 57, 563, 1986.

28. Armstrong, P.W., Stopps, T.P., Ford, S.E., and deBold, A.J., Rapid ventricular pacing in the dog: Pathophysiologic studies of heart failure, <u>Circulation</u>, 74, 1075, 1986.

29. Moe, G.W., Stopps, T.P., Howard, R.J., and Armstrong, P.W., Early recovery from heart failure: Insights into the pathogenesis of experimental chronic pacing-induced heart failure, <u>J. Lab. Clin. Med.</u>, 112, 426, 1988.

30. Riegger, G.A.J., Liebau, G., Holzschuh, M., Wilkowski, D., Steilner, H., and Kochsiek, K., Role of the renin-angiotensin system in the development of congestive heart failure in the dog as assessed by chronic converting-enzyme blockade, <u>Am. J. Cardiol.</u>, 53, 614, 1984.

31. Moe, G.W., Stopps, T.P., Angus, C., Forster, C., deBold, A.J., and Armstrong, P.W., Alterations in serum sodium in relation to atrial natriuretic factor and other neuroendocrine variables in experimental pacing-induced heart failure, <u>J. Am. Coll. Cardiol.</u>, 13, 173, 1989.

32. Coleman, H.N., Taylor, R.R., Pool, P.E., Whipple, G.H., Covell, J.W., Ross, J., and Braunwald, E., Congestive heart failure following chronic tachycardia, <u>Am. Heart J.</u>, 81, 790, 1971.

33. Dibner-Dunlap, M.E., and Thames, M.D., Baroreflex control of renal sympathetic nerve activity is preserved in heart failure despite reduced arterial baroceptor sensitivity, <u>Circ. Res.</u>, 65, 1526, 1989.

34. Pierpont, M.E.M., Judd, D., Foker, J.E., and Pierpont, G.L., Myocardial carnitine depletion associated with rapid ventricular pacing, <u>J. Am. Coll. Cardiol.</u>, 15, 10A, 1990.

DISCUSSION OF THE PAPER

G. HUG *(Cincinnati, OH)*: If this is a transport defect in your patients then why is the blood carnitine concentration so low?

PIERPONT: The transport defect is also manifest in the kidneys besides muscle and fibroblasts. Tubular reabsorption is deficient and results in loss of carnitine in the urine thus lowering the plasma carnitine (Treem et al, N Eng J Med 319:1331, 1988; Tein et al, Pediatr Res 28:247, 1990).

C. HOPPEL *(Cleveland, OH)*: Have you measured the decay of plasma carnitine during the dosing interval? A report in European J. Pediatrics used large (1800 mg/kg) doses of carnitine to increase the concentration gradient for driving diffusion into tissues.

PIERPONT: No, we have not yet measured plasma carnitine decay during the dosing interval in the two children with membrane carnitine transport defect. We are planning to do this. So far, we have not found that such doses of carnitine as were described in Rodrigues-Pereira et al (Eur J Pediatr 1487:193, 1988) are necessary to maintain the plasma carnitine of our two patients.

P. SUPPLE *(Downers Grove, IL)*: Is carnitine useful for VSD causing cardiac failure?

PIERPONT: Carnitine deficiency has not been reported in VSD induced heart failure. No treatment trials of carnitine in patients with symptomatic VSD have been published.

F. DiLISA *(Padova, Italy)*: What is the reason for the difference in carnitine content between biopsy and autopsy samples? Can they be explained on the basis of hypoxic conditions occurring after death?

PIERPONT: There are several reasons why differences in carnitine content can occur between cardiac biopsy and autopsy samples. The first consideration is whether biopsies are adequate samples, since they are not transmural and they are non-homogenous due to patchy distributions of fibrosis. Autopsy samples tend to be larger and be taken in regions where fibrosis is less. All of our cardiac carnitine measurements are quantitated per mg of non-collagenous protein to eliminate as much as possible errors caused by fibrosis. A second consideration in the interpretation of biopsy and autopsy specimens is the general health and nutritional state changes which the patient may undergo in the time just prior to death. If a lengthy period of poor nutrition occurs, there may be marked alterations in cardiac carnitine at autopsy for other than cardiac reasons. It does not seem likely that the differences between biopsy and autopsy carnitine can be explained by hypoxic conditions. In ischemia and hypoxia, elevated levels of cardiac acylcarnitines have been described (Shug and Paulson, in <u>Myocardial Ischemia and Lipid Metabolism</u> Plenum, 1984, p. 203.) and elevated acylcarnitines are not found in autopsy samples.

CARNITINE DEFICIENCY AND THE DIABETIC HEART

Dennis J. Paulson, Mohammed Sanjak, and Austin L. Shug
Department of Physiology, Chicago College of Osteopathic
Medicine and Department of Neurology, University of Wisconsin

I. CONDITIONS ASSOCIATED WITH MYOCARDIAL CARNITINE DEFICIENCY

L-Carnitine is the requisite carrier of long-chain fatty acids into mitochondria and may also participate in other aspects of cardiovascular energy metabolism such as modulation of the intramitochondrial acyl CoA : free CoA ratio and the removal of excess organic acids from the body(1). Long et al.(2) established that a relationship exists between tissue carnitine content and the optimal rate of long chain fatty acid oxidation. Since free fatty acids (FFA) are the preferred and predominant metabolic substrate used for energy production by the heart, alterations in myocardial levels of carnitine may have adverse effects on myocardial metabolism and contractile performance(1). A growing number of clinical reports have demonstrated a close association between primary carnitine deficiency and cardiomyopathy(3-5). In many cases, carnitine supplementation has been shown to improve mechanical function of these hearts. Because of this apparent link between carnitine content and functional performance of the myocardium, the role of carnitine in myocardial metabolism and contractile performance has become an area of intense interest. While the data on primarily genetic carnitine deficiency is interesting, this condition is clinically rare and not all cases are associated with cardiomyopathy(1). Carnitine deficiency can be acquired secondary to other diseases such as organic aciduria(6), renal disease(1), Reyes Syndrome(7), aging(8), E. Coli sepsis(9), diphtheria toxin(10), and diabetes(4,11-18). Often these conditions are associated with impaired cardiac contractile performance. However, the magnitude of the carnitine deficiency found with these conditions are relatively modest compared to that produced by primary carnitine deficiency. For example, the reported range for the myocardial L-carnitine deficiency found in diabetic animals ranges from 15 to 38 percent below normal values (11-18). Thus, the role of these modest carnitine deficiencies in the cardiac abnormalities associated with other diseases remains uncertain.

The purpose of this review is to discuss the relationship between myocardial L-carnitine content and cardiac contractile function. The role of the relatively small myocardial L-carnitine deficiency associated with diabetes mellitus as a contributing factor to the cardiac depression found with this disease is examined. New data on these subjects are presented.

II. D-CARNITINE-INDUCED L-CARNITINE DEFICIENCY

A precise quantitation of the relationship between myocardial L-carnitine content and contractile performance has been difficult to assess because of the lack of a good experimental model. One of the few experimental models of L-carnitine deficiency was developed by Paulson and Shug (19). L-Carnitine deficiency was produced in rats by intraperitoneal injection of D-carnitine, the inactive isomer. This treatment decreased myocardial L-carnitine content by 67%. When these hearts were excised, mounted on an isolated working heart perfusion apparatus and perfused with only palmitate as an exogenous substrate, they were unable to maintain normal levels of cardiac contractile performance while control hearts showed no such depression. This observation suggests that both fatty acid oxidation and heart contractile function were impaired by D-carnitine treatment. We hypothesized that the detrimental effects were due to a L-carnitine deficiency, however, since D-carnitine can inhibit acyltransferase activities this too may be another mechanism.

To examine this relationship further, we have attempted to determine whether the D-carnitine induced cardiac depression was due to an inability of the heart to utilize fatty acids as a substrate or rather an intrinsic depression in cardiac contractile performance. We also attempted to determine whether an elevation of myocardial L-carnitine content would enhance cardiac contractile

Table 1. Effects of 3 weeks of L-or D-carnitine injection (750 mg/kg/day) on myocardial L-carnitine content in the rat.

Group (n)	Total carnitine	Free carnitine (nmol/g dry wt)	Long-chain acylcarnitine	Short-chain acylcarnitine
Control saline (6)	5298 ± 625	2422 ± 509	1313 ± 280	1614 ± 334
L-Carnitine (6)	$7813 \pm 1181^{+}$	2990 ± 1117	$2267 \pm 387^{+}$	$2661 \pm 941^{+}$
D-Carnitine (6)	$3466 \pm 393^{\S+}$	$1016 \pm 128^{\S+}$	$1151 \pm 284^{\S}$	996 ± 165

* All values are mean $\pm$ SD for n shown in parentheses.
[+] Significantly different from saline control group (p < 0.05).
[§] Significantly different from L-carnitine group (p < 0.05).

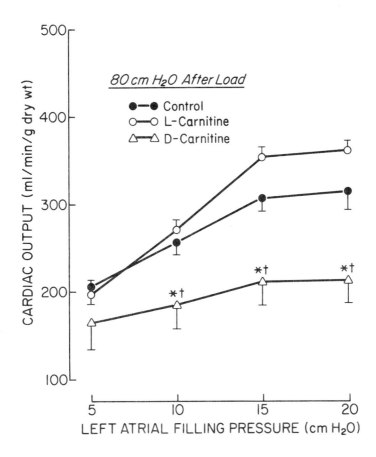

Figure 1: Effects of 3 weeks of L- or D-carnitine injection (750 mg/kg) on cardiac output curves of isolated perfused working rat hearts. The perfusion medium contained 1.2 mM palmitate and 5.5 mM glucose. All values are mean ± S.E.M. for n of 6 hearts. * $p < 0.05$ significantly different from the control group. + $p < 0.05$ significantly different from the L-carnitine group.

performance. Myocardial levels of L-carnitine were altered in rats by intraperitoneal injections of either L- or D-isomer of carnitine (750 mg/kg/day). A three week treatment period was used. At which point, in vitro cardiac pump performance was assessed in isolated perfused working hearts. Hearts were perfused with Krebs-Henseleit bicarbonate buffer gassed with 95% oxygen and 5% carbon dioxide and containing both palmitate (1.2 mM) and glucose (5.5 mM) as substrates. Left atrial filling pressure was sequentially adjusted to 5, 10, 15, 20 cm water at 5 minute intervals by varying the reservoir height. The effects on cardiac output were measured by timed collection. Hearts were paced at 315 beats/min. Our results demonstrated that the sum of free, long-chain and short-chain acyl-carnitine was increased by 34% after L-carnitine treatment while the D-carnitine injected animal showed a 54% decrease in the level of carnitine and its esters (Table 1). Even though glucose was present in the perfusion medium, the cardiac output curve of the D-carnitine supplemented group showed a significant downward shift to the right, thus, confirming that cardiac contractile performance was depressed in hearts with a L-carnitine deficiency (Figure 1). In contrast, the L-carnitine supplemented group had slightly elevated cardiac output values at the higher work loads but this increase was not statistically significant. These results support the hypothesis that L-carnitine deficiency can produce impaired cardiac contractile performance. L-carnitine supplementation of normal rats does not appear to have a beneficial effect on cardiac contractile function.

III. DIABETES-INDUCED MYOCARDIAL L-CARNITINE DEFICIENCY

A decrease in cardiac contractility in diabetic animals and man has been well established(20-22). Hearts from diabetic animals exhibit impaired functional performance in response to increased preloads(23-25) or β-adrenergic(26) stimulation. Hearts from diabetic animals show a decreased ability to recovery cardiac contractile function after anoxia and ischemia(27-30). A faster rate of ventricular failure in perfused ischemic hearts has also been observed(13). The metabolic mechanisms responsible for these defects in cardiac contractile performance are uncertain.

A number of studies have demonstrated that levels of carnitine are reduced by 15 to 38 % in hearts from diabetic animals depending upon the severity and duration of the diabetic state(11-15,17-18). It has also been observed that young insulin-treated diabetic patients have lower total and free serum carnitine(4,16). Although not studied in detail this L-carnitine deficiency is most likely the result of depressed myocardial uptake secondary to reduced blood carnitine content, however there is also evidence for defective myocardial L-carnitine transport and/or decrease renal reabsorption(18). The question we addressed was whether the much smaller myocardial L-carnitine deficiency found in diabetic animals was a contributing factor to the diabetes-induced cardiac depression.

A. EXERCISE TRAINING

As shown in Figure 2, it is possible to increase myocardial carnitine content of normal rat by treadmill exercise training. A significant and progressive increase in myocardial total L-carnitine content was produced after 6 and 12 weeks of treadmill exercise training. The increase was primarily in the short chain acyl fraction, but long chain acyl carnitine did show a smaller but significant increase.

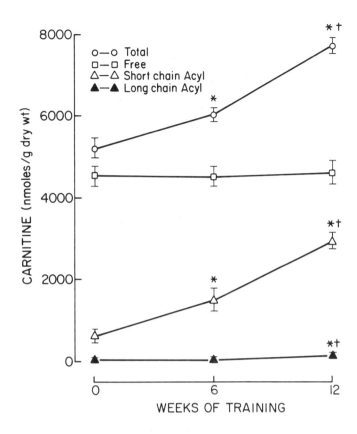

Figure 2: Effect of exercise training on myocardial L-carnitine content. Rats were trained on a treadmill for either 6 or 12 weeks at 27 m/min, 5% grade, 5 days per week. All values are mean $\pm$ S.E.M. for n of 8. * $p < 0.05$ significantly different from initial control value. + $p < 0.05$ significantly different from the 6 weeks value.

Based upon these findings we hypothesized that it may be possible to prevent the diabetes-induced myocardial L-carnitine deficiency by exercise training(24). The effects of exercise training of streptozotocin-induced diabetic rats on myocardial contractile performance and carnitine content were examined. Four groups were studied: sedentary control, trained control, sedentary diabetic and trained diabetic. The exercised trained diabetic rats were adapted to the treadmill prior to the induction of diabetes with i.v. streptozotocin (50 mg/kg). Thereafter the duration, speed and grade were progressively increased until the trained rats could run for 60 min., 27 m/min., 5% grade, 7 days per week for an eight week duration. At this point, cardiac output was measured in isolated working hearts perfused at various left atrial filling pressures and with buffer containing the concentrations of glucose and fatty acids found in poorly controlled diabetic plasma (22 mM glucose, 1.2 mM palmitate). Cardiac pump function was significantly depressed in hearts from the sedentary diabetic animals relative to sedentary controls rats. Myocardial carnitine content was also decreased by 22%. Exercise treadmill training prevented this cardiac depression of diabetic animals. This beneficial effects of exercise training were associated with an increase in myocardial total carnitine content of diabetic hearts back to control levels. Exercise training had no effect on cardiac performance of control animals, but did cause an increase in cardiac total L-carnitine content. While these findings suggest that there may be an association between the small L-carnitine deficiency and depressed cardiac pump performance of the diabetic heart, it is impossible to determine whether there is a cause and effect relationship. Exercise training also lowered plasma lipids in this study which may be an important beneficial action.

B. L-CARNITINE THERAPY

Previously, we have shown that intraperitoneal injection of L-carnitine (750 mg/kg) into two-week diabetic rats will prevent the diabetes-induced myocardial L-carnitine deficiency(12). Isolated perfused working hearts from these animals were subjected to various periods of ischemia followed by reperfusion. Hearts from the L-carnitine supplemented diabetic rats exhibited significantly enhanced recovery of cardiac contractile function. L-Carnitine treatment also lowered serum free fatty acids and cholesterol concentrations, but increased ketones levels. These results suggest that L-carnitine supplementation may prevent the diabetes-induced L-carnitine deficiency and may be beneficial to the diabetic heart. However, the duration of treatment (2 weeks) was relatively short in this study.

Rodrigues et al. (25,31) showed that chronic intraperitoneal injection of L-carnitine (3 g/kg/day) for 6 weeks prevented the onset of heart dysfunction in chronically diabetic rats. This improvement with L-carnitine supplementation was associated with an increase in myocardial free carnitine content and a lowering of plasma lipids. Subsequently, these authors used a lower dose of intraperitoneal L-carnitine (500 mg/kg/day) and were still able to demonstrate a similar beneficial effect of L-carnitine. In addition, these authors showed that the injection of L-carnitine for 2 weeks to 6 week diabetic rats was able to partially reversed the adverse effects of chronic diabetes on the heart.

We have also obtained evidence that chronic intraperitoneal injection of L-carnitine to diabetic rats prevents cardiac depression and lowers plasma lipids (Figure 3). Diabetes was induced by i.v. streptozotocin (55 mg/kg). The control and diabetic rats were divided into 2 subgroups: half received intraperitoneal injected L-carnitine (750 mg/kg/day) while the other half received saline. After 8 weeks, cardiac contractile performance was assessed in isolated perfused working hearts as described above. Saline-treated diabetic

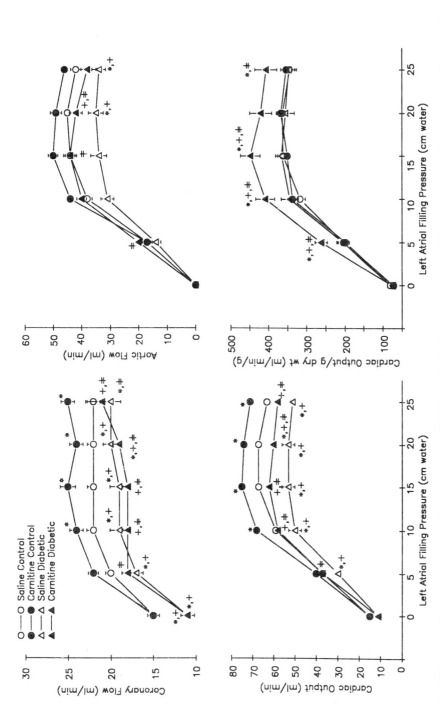

Figure 3: Effects of intraperitoneal L-carnitine (750mg/kg) therapy on cardiac pump performance of control and diabetic rats. All values are mean ± S.E.M. for n = 15 to 20. *p < 0.05 significantly different from saline control. +p < 0.05 significantly different from carnitine control. ‡p < 0.05 significantly different from saline diabetic.

rats showed a depression in aortic flow and cardiac output relative to saline treated control rats. The L-carnitine treated diabetic rats showed a significant improvement in cardiac pump performance. L-carnitine therapy had no effect in control rats. This beneficial effect of L-carnitine was associated with a lowering of plasma triacylglycerol concentration (data not shown).

While these data are interesting, intraperitoneal injections of L-carnitine into diabetic patients is not a realistic mode of therapy. Recently, Rodriques et al. (32) showed that oral L-carnitine therapy (200 mg/kg/day) for 6 weeks was ineffective in preventing the diabetes-induced cardiac depression of diabetic rats. This treatment also had no effect on plasma carnitine and lipid levels. However, it should be pointed out that the rats in this study were severely diabetic and were not treated with insulin. This situation is usually not present with human diabetes. The question remains whether oral L-carnitine supplementation to poorly controlled insulin treated diabetic animals will be effective in preventing the L-carnitine deficiency and cardiac depression.

To address this issue, we induced severe diabetes mellitus by using a relatively high dose of streptozotocin, 75 mg/kg. In order to prevent ketoacidosis and death, the diabetic rats were maintained on a relatively low daily dose of insulin, 0.3 units/100g body weight. In addition, another low dose insulin treated diabetic group was supplemented with 1 mg of L-carnitine/ml added to drinking water. Treatment was initiated 72 hours after the streptozotocin injection and was maintained for 10 weeks. Based upon water intake, the average dose of L-carnitine received by each rat over the 10 week treatment period was determined to be 486 mg/kg/day. This dose is considerably higher than that used by Rodrigues et al.(32). Cardiac pressure function was measured in the isolated perfused hearts using a fluid filled balloon inserted into the left ventricle. Two pressure-volume curves were measured for each heart. The first curve was derived when the hearts were perfused with buffer containing 3% albumin bound to either 0.4 or 1.2 mM palmitate plus 5.5 mM glucose. The second curve was obtained after the perfusion medium was switched to buffer containing the other concentration of palmitate and 5.5 mM glucose.

Figure 4 shows the changes in left intraventricular pressure development in response to increasing balloon volumes (preload) of the hearts perfused with medium containing normal levels of glucose (5.5 mM) and elevated concentration of palmitate (1.2 mM). The hearts from the low dose insulin-dependent diabetic group showed a significant depression in both positive and negative dP/dt relative to the control group. This depression in cardiac contractile performance was much less and only evident at high preloads when these hearts were perfused with normal levels of glucose (5.5 mM) and palmitate (0.4 mM) (data not shown). Oral L-carnitine administration to similarly insulin treated diabetic rats prevented this depression in these cardiac contractile parameters. Left intraventricular systolic pressure was elevated in

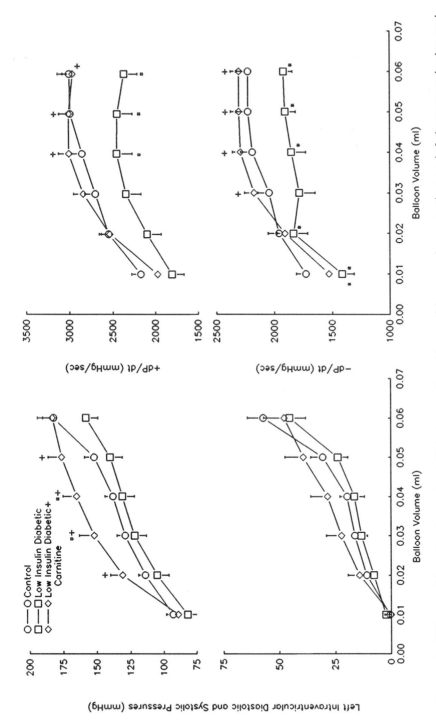

Figure 4: Effects of insulin-dependent diabetes and oral L-carnitine supplementation on left intraventricular end-diastolic and systolic pressure and positive and negative dP/dt of isolated hearts perfused with normal levels of glucose(5.5 mM), but elevated concentration of palmitate(1.2 mM). Hearts were subjected to progressive increases in left intraventricular end diastolic volume using a fluid filled balloon inserted into the left ventricle. All values are mean ± S.E.M. for n= 11 to 26. *P<0.05 significantly different from Control. +P<0.05 significantly different from Low Insulin Diabetic.

the L-carnitine supplemented group, suggesting that there may have been some inotropic effect of L-carnitine.

After perfusion, the hearts were analyzed for total L-carnitine content (Figure 5). Myocardial total carnitine content was significantly reduced by 18% in the low dose insulin treated diabetic group relative to the control group. Oral L-carnitine supplementation to diabetic rats prevented this loss. The decrease in cardiac total carnitine content of the low dose insulin treated diabetic group was associated with reduced plasma levels of both acid-soluble and total carnitine. The acid-insoluble fraction was increased in the low dose insulin diabetic group relative to control (data not shown). L-carnitine administration to the low dose insulin treated diabetic rats raised plasma levels to within the control range.

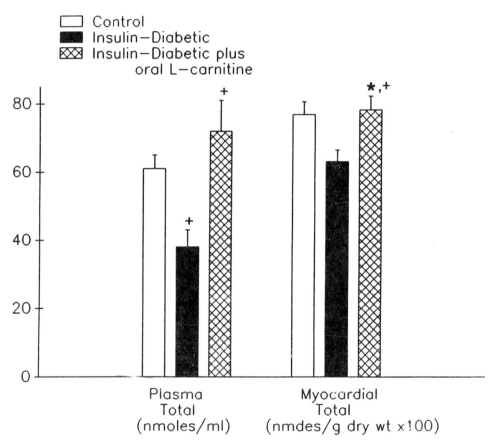

Figure 5: Effect of oral L-carnitine supplementation on plasma total L-carnitine and myocardial total L-carnitine content of poorly controlled insulin treated diabetic rats. All values are mean $\pm$ S.E.M. for n = 12 to 19. *P < 0.05 significantly different from control. $^{+}$P < 0.05 significantly different from insulin-diabetic.

The present study demonstrated that the rates of pressure development and relaxation were depressed only at very high preloads in poorly controlled streptozotocin-induced insulin-dependent diabetic rats perfused with normal levels of exogenous glucose and fatty acids, but this cardiac depression becamed more pronounced when the concentration of exogenous palmitate was elevated to levels found with poorly controlled diabetes. The glucose concentration of the perfusion medium was maintained at normal levels throughout the study. As expected, elevated exogenous palmitate had no effect on cardiac performance of control rats. Previously, we have shown that the magnitude of the cardiac output depression of noninsulin-treated diabetic rats was apparently greater when the hearts were perfused with elevated glucose and palmitate (24). However, we could not determine whether this depression was due to an elevation of palmitate, glucose or both. Since in the present study, cardiac performance assessments with high and low exogenous palmitate levels were performed in the same heart and only palmitate concentration was elevated, the results indicated that the further depression in cardiac performance found in poorly controlled diabetes was due to elevated exogenous palmitate. Oral L-carnitine supplementation prevented the myocardial L-carnitine deficiency of diabetic rats. The cardiac depression observed when diabetic hearts were perfused with high levels of exogenous free fatty acids was not found in hearts from the L-carnitine treated diabetic rats.

Oral L-carnitine supplementation to diabetic rats also has a lipid lowering action (Figure 6). The insulin-treated diabetic rats receiving low dose insulin plus oral L-carnitine exhibited significantly reduced plasma concentration triglycerides but not cholesterol. Overall glycemic control was not affected as indicated by the level of blood glycosylated hemoglobin.

In contrast to the results of Rodrigues et al. (32), oral L-carnitine therapy was effective in the present in preventing cardiac depression and the loss of L-carnitine from the heart. This discrepancy is most likely due to the fact that we used a higher dose of L-carnitine and that it was given in combination with insulin therapy.

III. SUMMARY AND CONCLUSIONS

These findings provide further evidence that myocardial L-carnitine deficiency can adversely affect cardiac contractile performance. Even the relatively small reduction in myocardial L-carnitine content found in diabetic animals appears to contributed to the depression in cardiac contractile performance. In the present study, prevention of the myocardial L-carnitine deficiency by exercise training or supplementation with either oral or intraperitoneal injected L-carnitine, resulted in a significant improvement in cardiac contractile performance. In addition, both exercise training and

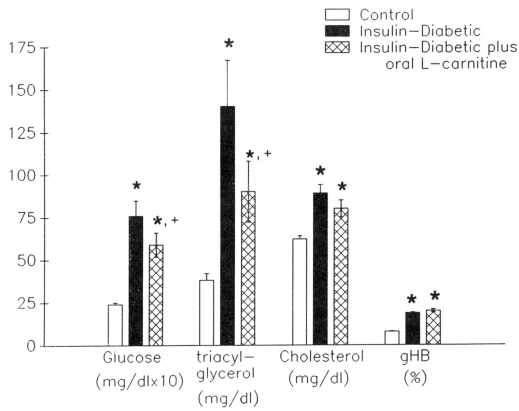

Figure 6: Effect of oral L-carnitine supplementation on plasma glucose, triacylglycerol, cholesterol and blood glycosylated hemoglobin of poorly controlled insulin treated diabetic rats. All values are mean ± S.E.M. for n = 12 to 19. *P < 0.05 significantly different from control. +P < 0.05 significantly different from insulin-diabetic.

L-carnitine supplementation significantly decreased plasma triacylglycerol concentration of diabetic rats.

Precisely how the small L-carnitine deficiency found in the diabetic heart contributes to cardiac contractile performance is uncertain. The lack of adequate L-carnitine stores may interfere with the uptake of long chain fatty acids by the mitochondria, causing a buildup of tissue free fatty acid and cytoplasmic long chain acyl CoA with a commensurate decrease of free CoA. A shift in the flux of fatty acid molecules into triacylglycerol pool of the heart may then occur. One or more of these metabolic alterations may result in impaired contractile performance of the diabetic heart. For example, high tissue levels of these lipid intermediates have been shown to specifically inhibit key metabolic processes such as sarcoplasmic reticulum Ca^{++}-ATPase(33), sarcolemmal Na^+,K^+-ATPase(34) and mitochondrial adenine nucleotide translocator(35). The accumulation of lipids may also alter membrane structure due to nonspecific detergent actions(36).

The beneficial effects of exercise training or L-carnitine administration may have been due to the lipid lowering effect of these treatment. Since lowering of plasma lipids by lipid lowering drugs such as hydralazine(37) have also been shown to prevent the cardiac depression found in the streptozotocin-induced noninsulin-dependent diabetic rat, the beneficial effects of L-carnitine may have been mediated through this mechanism. This hypothesis is supported by the study of Dillman (38) who found that the administration of methyl palmoxirate, a carnitine palmitolyl-transferase I inhibitor, for 4 weeks to diabetic rats prevented the depression of myosin ATPase activity and restored the isoenzyme pattern. Tahiliani and McNeill (39) showed that a combination of methyl palmoxirate and thyroid hormone to diabetic rats prevented the cardiac depression. These results suggest that long term elevation of exogenous lipids may cause alteration in gene expression for cardiac contractile protein which may lead to altered cardiac contractility. By lowering plasma lipids, exercise training and L-carnitine administration may have prevented these metabolic and functional changes.

Acknowledgments

L-carnitine was generously donated by Sigma Tau Pharmaceuticals, Rome, Italy. This work was supported from a Feasibility Grant from The American Diabetes Association, National Institute of Diabetes and Kidney Diseases Grant DK39200 and Sigma Tau Pharmaceuticals.

References

1. Rebouche, C.J. and Paulson, D.J. Carnitine metabolism and functions in humans. Ann. Rev. Nutr.,6,,41, 1986.
2. Long, C.S., Haller, R.G., Foster, D.W. and McGarry, J.D. Kinetics of carnitine dependent fatty acid oxidation: implication for human carnitine deficiency. Neurology,32,,663, 1982.
3. Tripp, M.E., Katcher, M.L., Peters, H.A., Gilbert, E.F., Hodach, R.J. and Shug, A.L. Systemic carnitine deficiency presenting as familial endocardial fibroelastosis. N. Engl. J. Med.,305,,385, 1981.
4. Winter, S.C., Simon, M., Zorn, E.M., et al. Relative carnitine insufficiency in children with type I diabetes mellitus. Am.J.Dis.Child.,143,,1337, 1989.

5. Waber, L.J., Valle, D., Neill, C., DiMauro, S. and Shug, A. Carnitine deficiency presenting as familial cardiomyopathy: A treatable defect in carnitine transport. <u>J. Pediat.</u>,101,,700, 1982.

6. Roe, C.R., Millington, D.S., Maltby, D.A., Kahler, S.G. and Bohan, T.P. L-carnitine therapy in isovaleric acidemia. <u>J. Clin. Invest.</u>,74,,2290, 1984.

7. Chapoy, P.R., Angelini, C., Brown, W.J., Stiff, J.E., Shug, A.L. and Cederbaum, S.D. Systemic carnitine deficiency-a treatable inherited lipid-storage disease presenting as Reye's Syndrome. <u>N. Engl. J. Med.</u>,303,,1389, 1980.

8. Hansford, R. Lipid oxidation by heart mitochondria from young adult and senescent rats. <u>Biochem. J.</u>,170,,285, 1978.

9. Lanza-Jacoby, S. and Reibel, D.K. Changes in tissue levels of carnitine during E. Coli sepsis in the rat. <u>Circ. Shock</u>,24,,29, 1988.

10. Bressler, R. and Wittels, B The effect of diphtheria toxin on carnitine metabolism in the heart. <u>Biochim. Biophys. Acta</u>,104,,39, 1965.

11. Fogle, P.J. and Bieber, L.L. Effects of streptozotocin on carnitine and carnitine acyltransferase in rat heart, liver and kidney. <u>Biochem. Med.</u>,22,,119, 1979.

12. Paulson, D.J., Schmidt, M.J., Traxler, J.S., Ramacci, M.T. and Shug, A.L. Improvement of myocardial function in diabetic rats after treatment with L-carnitine. <u>Metabolism</u>,33,,358, 1984.

13. Feuvray, D., Idell-Wenger, J.A. and Neely, J.R. Effects of ischemia on rat myocardial function and metabolism in diabetes. <u>Am. J. Physiol.</u>,44,,322, 1979.

14. Pearson, P.J. and Tubbs, P.K. Carnitine and derivatives in rat tissues. <u>Biochem. J.</u>,105,,953, 1967.

15. Bohmer, T., Norum, K.R. and Bremer, J. The relative amount of long-chain acylcarnitine, acetylcarnitine, and free carnitine in organs of rats in different nutritional states and with alloxan diabetes. <u>Biochim. Biophys. Acta</u>,125,,244, 1966.

16. Cederblad, G., Hermansson, G. and Ludvigsson, J. Plasma and urine carnitine in children with diabetes mellitus. <u>Clin. Chim. Acta</u>,125,,207, 1982.

17. Stearns, S.B. Carnitine content of skeletal muscle from diabetic and insulin-treated diabetic rats. <u>Biochem. Med.</u>,24,,33, 1980.

18. Brooks, S.D., Bahl, J.J. and Bressler, R. Carnitine in the streptozotocin-diabetic rat. <u>J. Nutr.</u>,1157,,1267, 1985.

19. Paulson, D.J. and Shug, A.L. Tissue specific depletion of L-carnitine in rat heart and skeletal muscle by D-carnitine. <u>Life Sci.</u>,28,,2931, 1981.

20. Regan, T.J., Wu, C.F., Yeh, C.K., Oldewurtel, H.A. and Haider, B. Myocardial composition and function in diabetes. <u>Circ. Res.</u>,49,,1268, 1981.

21. Tahiliani, A.G. and McNeill, J.H. Minireview, Diabetes-Induced Abnormalities in the Myocardium. Life Sci.,38,,959, 1986.

22. Kannel, W.B. and McGee, D.L. Diabetes and Cardiovascular Risk Factors: The Framingham Study. Circ.,59,,8, 1979.

23. Fein, F.S., Kornstein, L.B., Strobeck, J.E., Capasso, J.M. and Sonnenblick, E.H. Altered myocardial mechanics in diabetic rats. Circ. Res.,47,,922, 1980.

24. Paulson, D.J., Kopp, S.J., Peace, D.G. and Tow, J.P. Myocardial adaptation to endurance exercise training in diabetic rats. Am. J. Physiol.,252,,R1073, 1987.

25. Rodrigues, B., Ross, J.R., Farahbakshian, S. and McNeill, J.H. Effects of in vivo and in vitro treatment with L-carnitine on isolated hearts from chronically diabetic rats. Can.J.Physiol.Pharmacol.,68,,1085, 1990.

26. Paulson, D.J., Kopp, SJ., Tow, JP., Feliksik, JM. and Peace, DG. Impaired in vivo myocardial reactivity to norepinephrine in diabetic rats. Proc Soc Exp Biol Med,183,,186, 1986.

27. Hearse, D.J., Steward, D.A. and Green, D.G. Diabetes and the survival and recovery of the anoxic myocardium. J. Mol. Cell Cardiol.,7,,397, 1975.

28. Hearse, D.J., Steward, D.A. and Green, D.G. Myocardial susceptibility to ischemic damage: a comparative study of disease models of rats. Eur. J. Cardiol.,75,,437, 1978.

29. Ingebretsen, C.G., Moreau, P., Hawelu-Johnson, C. and Ingebretsen, W.R. Performance of diabetic rat hearts: effects of anoxia and increased work. Am. J. Physiol.,239,,H614, 1980.

30. Paulson, D.J., Kopp, S.J., Peace, D.G. and Tow, J.P. Improved postischemic recovery of cardiac pump function in exercised trained diabetic rats. J. Appl. Physiol.,65,,187, 1988.

31. Rodrigues, B., Xiang, H. and McNeill, J.H. Effect of L-carnitine treatment on lipid metabolism and cardiac performance in chronically diabetic rats. Diabetes,37,,1358, 1988.

32. Rodrigues, B., Seccombe, D. and McNeill, J.H. Lack of effect of oral L-carnitine treatment on lipid metabolism and cardiac function in chronically diabetic rats. Can.J.Physiol.Pharmacol.,68,,1601, 1990.

33. Lopaschuk, G.D., Tahiliani, G., Vadlamudi, R.V.S.V., Katz, S. and McNeill, J.H. Cardiac sarcoplasmic reticulum function in insulin- or carnitine-treated diabetic rats. Am. J. Physiol.,245,,H969, 1983.

34. Wood, J.M., Bush, B., Pitts, B.J.R. and Schwartz, A. Inhibition of bovine heart $Na+,K+$-ATPase by palmitylcarnitine and palmityl-CoA. Biochem. Biophys. Res. Commun.,74,,677, 1977.

35. Paulson, D.J. and Shug, A.L. Inhibition of the adenine nucleotide translocator by matrix-localized palmityl-CoA in rat heart mitochondria. Biochim. Biophys. Acta,766,,70, 1984.

36. Katz, A.M. and Messineo, F.C. Lipid-membrane interactions and the pathogenesis of ischemic damage in the myocardium. <u>Circ. Res.</u>,48,,1, 1981.

37. Rodrigues, B., Goyal, R.K. and McNeill, J.H. Effects of hydralazine on streptozotocin-induced diabetic rats: prevention of hyperlipidemia and improvement of cardiac function. <u>J. Pharmacol. Exp. Ther.</u>,237,,292, 1986.

38. Dillmann, W.H. Methyl palmoxirate increases calcium-myosin ATPase activity and changes myosin isoenzyme distribution in the diabetic rat. <u>Am. J. Physiol.</u>,148,,E602, 1985.

39. Tahiliani, A.G. and McNeill, J.H. Prevention of diabetes-induced myocardial dysfunction in rats by methyl palmoxirate and triiodothyronine treatment. <u>Can. J. Physiol. Pharm.</u>,63,,925, 1985.

BOTH CARNITINE AND CARNITINE PALMITOYLTRANSFERASE 1 INHIBITORS STIMULATE GLUCOSE OXIDATION IN ISOLATED HEARTS PERFUSED WITH HIGH CONCENTRATIONS OF FATTY ACIDS

Gary D. Lopaschuk, Tom L. Broderick, Arthur H. Quinney, and Maruf Saddik
Cardiovascular Disease Research Group
Lipid and Lipoprotein Research Group
423 Heritage Medical Research Bldg
The University of Alberta
Edmonton, Canada, T6G 2S2

I. INTRODUCTION

Increasing myocardial carnitine content can have a protective effect on ischemic myocardium[1-6] and can improve heart function in cardiomyopathies associated with diabetes and carnitine-deficiency[7-10]. Studies have also demonstrated that carnitine palmitoyltransferase 1 (CPT 1) inhibitors can have a protective effect on ischemic myocardium[11-15] and can also improve mechanical function in hearts from diabetic animals[15-17]. The mechanisms by which carnitine and CPT 1 inhibitors exert these effects has still not been completely delineated. The commonly proposed theories to explain the actions of carnitine, however, appear to contradict the commonly proposed mechanisms to explain the actions of CPT 1 inhibitors. Carnitine, by enhancing CPT 1 activity, has been suggested to enhance fatty acid oxidation and prevent the accumulation of potentially toxic levels of cytosolic long chain acyl CoA. The accumulation of long chain acyl CoA has been suggested to result in an inhibition of the mitochondrial ATP transporter[18]. In contrast, CPT 1 inhibitors are thought to inhibit fatty acid oxidation and prevent the accumulation of long chain acylcarnitine by inhibiting the transfer of acyl groups from long chain acyl CoA to long chain acylcarnitine[19,20]. This theoretically should actually result in an increase in cytosolic levels of long chain acyl CoA, which can inhibit the mitochondrial ATP transporter. In attempting to clarify this controversy, we directly determined the effects of carnitine and CPT 1 inhibitors on energy substrate utilization in isolated working hearts perfused with relevant concentrations of fatty acids. Based on direct measurements of glucose and palmitate oxidation, we propose that the beneficial effects of carnitine and CPT 1 inhibitors can both be explained by the effects of these agents on overcoming fatty acid inhibition of glucose oxidation.

II. METHODS

A. HEART PERFUSIONS

Adult male Sprague Dawley rats (200-250 g) were anesthetized with sodium pentobarbital (60 mg/kg). Hearts were quickly excised and cannulated as working hearts as described previously [13,14]. Hearts were perfused at an 11.5 mm Hg left atrial preload and an 80 mm Hg aortic afterload with Krebs-Henseleit buffer containing 11 mM glucose, 3% albumin, and 2.5 mM free Ca^{2+}. Hearts were perfused in the presence or absence of 1.2 mM palmitate, as indicated in the individual experiments. When used, palmitate was pre-bound to the albumin. The radioisotopes of glucose or palmitate that were used are also indicated in the individual experiments.

In experiments in which myocardial carnitine content was increased, hearts were perfused with 10 mM carnitine for a 60 minute period prior to the measurement of glucose oxidation or glycolysis.

B. MEASUREMENT OF GLYCOLYSIS, GLUCOSE OXIDATION, AND PALMITATE OXIDATION

To measure palmitate oxidation, hearts were perfused with 11 mM glucose and 1.2 mM $(1-{}^{14}C)$-palmitate. Steady state exogenous palmitate oxidation was determined by quantitatively measuring $^{14}CO_2$ production by hearts, as described in detail previously[13,14]. Perfusate and gaseous samples were collected at 10 minute intervals over a 60 minute period for determination of $^{14}CO_2$ production.

In hearts used for glycolysis and glucose oxidation measurements, perfusate contained 11 mM $(2-{}^3H/U-{}^{14}C)$-glucose. Quantitative ^3H_2O production was used to measure steady state glycolytic rates (^3H_2O is liberated at the phosphoglucoisomerase step of glycolysis), while quantitative $^{14}CO_2$ production was used to measure glucose oxidation ($^{14}CO_2$ is liberated at the level of PDH and in the TCA cycle). $^{14}CO_2$ production was determined using the same methods described above for palmitate oxidation. ^3H_2O production was measured as described previously[21,22].

Data are presented as the mean $\pm$ S.E of the mean. The students t-test was used to determine statistical significance between sample populations. A value of $p < 0.05$ was regarded as significant.

III. RESULTS AND DISCUSSION

A. INTERACTION OF FATTY ACIDS WITH MYOCARDIAL GLUCOSE USE

A complex interaction exists between fatty acid and glucose utilization in the intact heart (see 23, 24, and 25 for review). A dramatic example of this regulation can be seen if circulating fatty acid concentrations are elevated. Myocardial fatty acid oxidation rates increase with a concomitant decrease in both glycolysis and glucose oxidation. This is illustrated in Table 1, in which steady state glycolysis, glucose oxidation, and palmitate oxidation rates were measured in isolated working hearts. Paralleling our previous observations[21,22] and those of Kobyashi and Neely [26], we observed that in the absence of added fatty acids, glycolytic rates are approximately 2-fold greater than the rates of glucose oxidation. If 1.2 mM palmitate was present in the perfusate, both glycolysis and glucose oxidation rates decreased. The decrease in glucose oxidation, however, was much more dramatic than the decrease in glycolysis. As a result, glycolytic rates were found to be approximately 13-fold greater than glucose oxidation rates. This data confirms our previous data that fatty acids are much more potent inhibitors of glucose oxidation than glycolysis.

TABLE 1

Glycolysis, Glucose Oxidation, and Palmitate Oxidation In Isolated Working Rat Hearts

Perfusion Conditions	Glycolysis (nmol ^{3}H-glucose/ g dry wt·min)	Glucose Oxidation (nmol ^{14}C-glucose/ g dry wt·min)	Palmitate Oxidation (nmol ^{3}H-palmitate/ g dry wt·min)
-11mM glucose (n=6)	2698 ± 461	1257 ± 158	/
-11mM glucose, 1.2mM palmitate (n=5)	1678 ± 280	124 ± 26*	633 ± 60

Data are the mean ± S.E.M of a number of hearts indicated in brackets. Hearts were perfused in the working mode for 1 hour with 11 mM [2-^{3}H/U-^{14}C] glucose ± 1.2 mM palmitate. Glycolytic rates were determined by measuring perfusate ^{3}H$_2$O levels while glucose oxidation rates were determined by measuring ^{14}CO$_2$ production. Another series of hearts were perfused with 1.2 mM [9,10-^{3}H] palmitate + 11 mM glucose, and palmitate oxidation rates were determined by measuring perfusate ^{3}H$_2$O levels.
*, significantly different from hearts perfused in the absence of fat.

The decrease in glycolysis occurs primarily as a result of an increase in cytosolic citrate concentrations. Extensive studies by Randle's laboratory (see 24 for review) have demonstrated that the major determinant of glucose oxidation in aerobically perfused hearts is flux through pyruvate dehydrogenase (PDH). During starvation or in diabetes, flux through PDH is markedly decreased, while following a carbohydrate rich diet the activity of PDH increases. Regulation of pyruvate dehydrogenase occurs via a phosphorylation-dephosphorylation cycle which is subject to both intrinsic and extrinsic effectors (see 24 and 25 for reviews). Phosphorylation of the PDH complex results in a decrease in its activity, while dephosphorylation increases its activity. Increases in intramitochondrial NADH/NAD⁺, acetyl-CoA/CoA, and ATP/ADP ratios will increase phosphorylation of PDH resulting in a decrease in activity of the PDH enzyme complex. One of the effects of increasing fatty acid oxidation is to increase these ratios, particularly the intramitochondrial acetyl-CoA/CoA ratio, thereby decreasing PDH activity. This is probably the key factor which results in the dramatic decrease in glucose oxidation observed in hearts perfused with high concentrations of fatty acids. This effect of high levels of fatty acids is shown diagramatically in Figure 1.

FIGURE 1

Mechanism by which fatty acids inhibit glucose oxidation

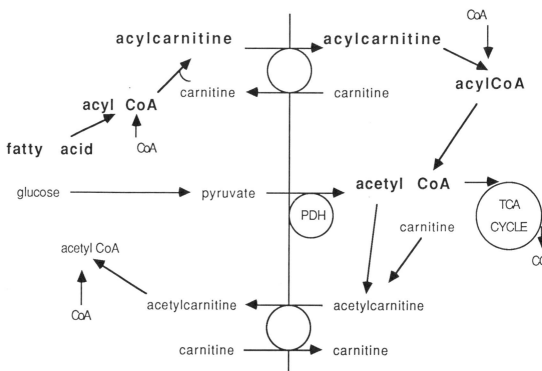

High levels of fatty acids results in an increase in the intermediates involved in fatty acid oxidation (indicated in bold lettering). The accumulation of intramitochondrial acetyl CoA inhibits pyruvate dehydrogenase activity (PDH), resulting in a decreased flux of pyruvate through PDH. The net effect is a decrease in CO_2 production from glucose.
TCA, tricarboxylic acid cycle; CoA, coenzyme A.

Although fatty acid oxidation is the primary source of ATP production in the aerobic heart (see 23 for review), the amount of fatty acid utilized varies, depending primarily on the workload of the heart, and the concentration of circulating fatty acids. We recently performed a series of experiments in which the relative proportion of ATP production from glycolysis, glucose oxidation, exogenous fatty acid oxidation, and from endogenous triglycerides was determined[22]. In the presence of normal levels of fatty acids (0.4 mM palmitate), 76% of total myocardial ATP production in isolated working hearts is obtained from fatty acids. In hearts perfused with high concentrations of fatty acids, 92% of total ATP production is obtained from fatty acid oxidation[22]. Even if hearts are perfused in the absence of added fatty acids, 42% of steady state ATP production is derived from the oxidation of fatty acids originating from endogenous triglyceride hydrolysis. These results demonstrate the importance of fatty acids as an energy substrate, and suggest that contractile function of the heart cannot be maintained by glucose utilization alone.

B. EFFECTS OF CARNITINE PALMITOYLTRANSFERASE (CPT 1) INHIBITION ON GLUCOSE OXIDATION AND GLYCOLYSIS IN FATTY ACID PERFUSED HEARTS

CPT 1 inhibitors are a class of compounds that inhibit the activity of carnitine palmitoyltranferase 1, which is present on the outer side of the inner mitochondrial membrane. In isolated mitochondria, CPT 1 inhibitors will result in marked decreases in the rate of oxidation of free fatty acids or acyl CoA[19,20]. These compounds have been used in a number of intact heart or intact animal studies as agents to block fatty acid oxidation. Despite this, however, very few studies have actually directly measured the effects of CPT 1 inhibitors on fatty acid oxidation in the intact heart. In isolated working hearts perfused in the presence of high concentrations of fatty acid, we have been unable to observe any inhibition of fatty acid oxidation with the CPT 1 inhibitor, Etomoxir[13,14]. This is not surprising, since in these hearts over 90% of the steady state ATP production is originating from fatty acid oxidation[14,22]. As a result, in order to maintain ATP production (and presumably heart function), an inhibition of fatty acid oxidation would have to be accompanied by a parallel increase in ATP production from other sources. Etomoxir will increase glucose oxidation more than 100% in hearts perfused with high fat[13]. However, since glucose oxidation is providing less than 5% of total ATP production under these conditions, a dramatic inhibition of fatty acid oxidation would not be expected[22]. As a result, we suggest that in hearts perfused with high concentrations of fatty acids, the primary effect of CPT 1 inhibition is a decrease in the degree of fatty acid inhibition of glucose oxidation, rather than an actual large decrease in fatty acid oxidation. The effects of Etomoxir on glycolysis and glucose oxidation in hearts perfused in the presence of high fatty acid concentration is shown in Table 2. As expected, glycolytic rates were much greater in these hearts than glucose oxidation rates. Etomoxir, resulted in an increase in both glycolysis and glucose oxidation, although only the increase in glucose oxidation was significant . This effect of Etomoxir may occur due to a lowering of intramitochondrial acetyl CoA/CoA ratio (Figure 2). As discussed later, this increase in glucose oxidation may account for the observed beneficial effects of CPT 1 inhibitors in the diabetic and ischemic heart.

TABLE 2

Effect of the Carnitine Palmitoyltransferase 1 Inhibitor, Etomoxir, on Gycolysis and Glucose Oxidation in Isolated Working Hearts

Perfusion Conditions	Glycolysis (nmol ^{3}H-glucose/ g dry wt·min)	Glucose Oxidation (nmol ^{14}C-glucose/ g dry wt·min)
-no addition (n=5)	2180 ± 480	108 ± 35
-Etomoxir (n=6)	2690 ± 720	$216 \pm 46*$

Data are the mean ± S.E.M of a number of hearts indicated in brackets. Hearts were perfused in the working mode for 1 hour with 11 mM [2-^{3}H/U-^{14}C] glucose,1.2 mM palmitate. *, significantly different from those in hearts perfused in the absence of fat.

FIGURE 2

Proposed Mechanism By Which CPT 1 Inhibitors Increase Glucose Oxidation In Hearts Perfused With Fatty Acids

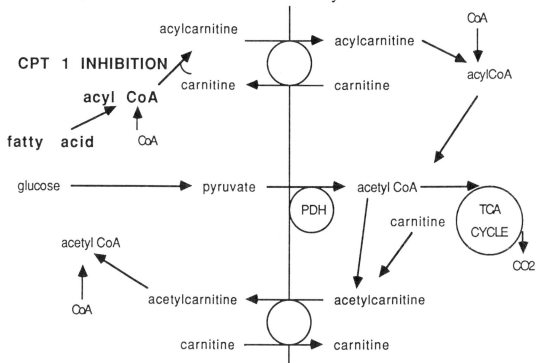

Inhibition of carnitine palmitoyltransferase 1 results in a lowering of fatty acid intermediates. A resultant decrease in intramitochondrial acetyl CoA levels results in an activation of PDH. The result is a stimulation of CO_2 production from glucose.
PDH, pyruvate dehydrogenase; TCA, tricarboxylic acid cycle; CoA, coenzyme A.

C. EFFECTS OF CARNITINE ON GLYCOLYSIS AND GLUCOSE OXIDATION IN HEARTS PERFUSED WITH HIGH CONCENTRATIONS OF FATTY ACIDS

In isolated mitochondrial preparations, carnitine lowers intramitochondrial acetyl CoA levels[27,28]. In heart mitochondria, carnitine increases CoA levels and reduces acetyl CoA levels, resulting in a 10- to 20-fold decrease in the ratio of acetyl CoA/CoA[29,31]. The change in this ratio of acetyl CoA/CoA correlates with efflux of acetylcarnitine from the mitochondria, which is consistent with the suggestion that carnitine increases the activity of carnitine acetyltransferase present on mitochondrial membranes.

As discussed earlier, the activity of the PDH complex is the major determinant of glucose oxidation in aerobically perfused hearts[24,25]. This complex is regulated by a phosphorylation-dephosphorylation cycle that is responsive to changes in the ratio of intramitochondrial acetyl CoA/CoA. As would be expected, therefore, carnitine by decreasing the intramitochondrial acetyl CoA/CoA ratio, stimulates PDH activity. (Figure 3).

FIGURE 3
Proposed Mechanism By Which Carnitine Increases Glucose Oxidation
In Hearts Perfused With High Concentrations of Fatty Acids

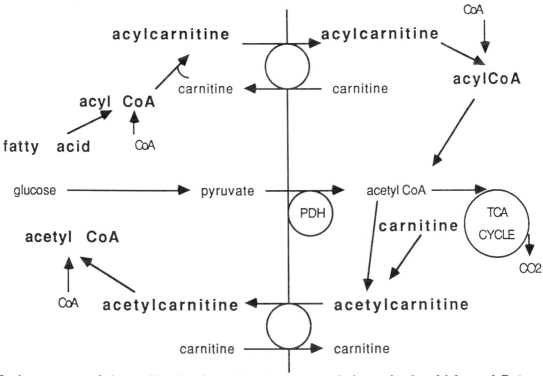

In the presence of elevated levels of carnitine, the increase in intramitochondrial acetyl CoA levels that is seen in the presence of high concentrations of fatty acids is prevented as a result of the transfer of the acetyl groups from acetyl CoA to acetylcarnitine. The decrease in acetyl CoA levels results in activation of PDH. This increases CO_2 production from glucose. PDH, pyruvate dehydrogenase; TCA, tricarboxylic acid cycle; CoA, coenzyme A.

We were interested in determining if the carnitine induced increase in PDH activity observed in isolated mitochondria, also occurred in the intact heart. To perform these experiments, we initially perfused hearts for a 60-minute period in the presence of 10 mM L-carnitine, followed by a 5 minute washout period to remove any extracellular carnitine. These conditions were chosen since it has previously been demonstrated that carnitine is transported across the sarcolemmal membrane in slow Na^+-dependent manner[32]. In our working hearts, a 60-minute perfusion period resulted in an increase in myocardial carnitine content from 4595 ± 258 to $11,122 \pm 720$ nmol/g dry wt. Glucose oxidation was subsequently measured in hearts perfused in the presence of 1.2 mM palmitate (Table 3). In hearts containing elevated levels of carnitine, a significant increase in glucose oxidation occurred. This effect of carnitine on glucose oxidation was accompanied by an increase in glycolysis.

TABLE 3

Effects of L-carnitine on Glycolysis and Glucose Oxidation in
Isolated Working Hearts Perfused with Fatty Acids

Perfusion Conditions	Glycolysis (nmol ^{3}H-glucose/ g dry wt·min)	Glucose Oxidation (nmol ^{14}C-glucose/ g dry wt·min)
-no addition (n=9)	2.91 ± 0.23	158.4 ± 21.4
-carnitine loaded (n=9)	$4.63 \pm 0.46*$	$454.1 \pm 85.3*$

Data are the mean ± S.E.M of a number of hearts indicated in brackets. Carnitine-loaded hearts were pre-perfused in the working mode for 1 hour with 10 mM carnitine. Glycolysis and glucose oxidation was measured by perfusing hearts with 11 mM [2-^{3}H/U-^{14}C] glucose and 1.2 mM palmitate. Glycolytic rates were determined by measuring perfusate ^{3}H$_2$O levels, while glucose oxidation rates were determined by measuring ^{14}CO$_2$ production. *, significantly different from those in hearts perfused in the absence of fat.

D. EFFECTS OF CPT 1 INHIBITORS AND CARNITINE ON ISCHEMIC AND DIABETIC HEARTS

A number of experimental studies have demonstrated that carnitine supplementation can have beneficial effects on ischemic and diabetic hearts[1-11]. Although the mechanisms by which this occurs are not known, it has been suggested that carnitine acts by stimulating fatty acid oxidation, or by decreasing the levels of long chain acyl CoA. Long chain acyl CoA is known to be a specific inhibitor of the ATP transporter located in the inner mitochondrial membrane[18]. However, in the intact heart a good correlation between accumulation of long chain acyl CoA and myocardial ATP levels has not been found[13,14,33]. Furthermore, fatty acid oxidation is not impaired in either the diabetic heart or in the reperfused ischemic heart[14,34-36]. In fact, in both of these conditions fatty acid oxidation can provide from 90 to 100% of myocardial ATP production[35,36].

CPT 1 inhibitors have also been shown to exert a beneficial effect on ischemic or diabetic hearts[11-15]. This has been thought to occur as a result of an inhibition of fatty acid oxidation and a decrease in the accumulation of long chain acylcarnitine within the myocardium[11,12,15]. However, we[13,14] and others[33] have been unable to show a good correlation between the beneficial effects of CPT 1 inhibitors and a lowering of long chain acylcarnitine levels. Similarly, we have also failed to find a correlation between the beneficial effects of CPT 1 inhibition and a decrease in palmitate oxidation.

IV. CONCLUSIONS

As discussed above, both carnitine and CPT 1 inhibition increase myocardial glucose oxidation in hearts perfused with high concentrations of fatty acids. It is our hypothesis that this increase in glucose oxidation may partly explain the beneficial effects of these agents during ischemia and in diabetes which has been observed in many studies. To date, we have yet to determine what effect acute carnitine loading of hearts has on glucose oxidation and function in diabetic hearts or in reperfused ischemic hearts. We have shown, however, that a good correlation exists between reperfusion recovery of function, and stimulation of glucose oxidation during reperfusion in hearts perfused with high concentrations of fatty acids[13,14,37]. Our results also suggest that stimulation of glucose oxidation can improve function in the diabetic heart[36,38]. We hypothesize, therefore, that stimulation of myocardial glucose oxidation by carnitine and CPT 1 inhibitors is part of the mechanism by which these agents exert their beneficial effect in diabetic and reperfused-ischemic hearts, despite what appears to be opposing actions on CPT 1 itself. Further studies are required to determine if this hypothesis is correct.

V. REFERENCES

1. Folts, J.D., Shug, A.L., Koke, J.R., and Bittar, N., Protection of the ischemic dog myocardium with carnitine, Am. J. Cardiol. 41,1209-1215, 1978.
2. Hulsmann, W.C., Dubelaar, M.L., Lamers, J.M.J., and Maccari, F., Protection by acyl-carnitine and phenylmethylsulphonyl fluoride of rat heart subjected to ischemia and reperfusion, Biochem. Biophys. Acta 847,62-66,1985.
3. Liedtke, A.J., and Nellis, S.H., Effect of carnitine in ischemic and fatty acid supplemented swine hearts, J. Clin. Invest. 64,440-447, 1979.

4. Paulson, D.J., and Shug, A.L., Effects of carnitine on the ischemic arrested heart, Basic Res. Cardiol. 241,H505-H512,1981.

5. Nagao, B., Kobayashi, A., and Yamazaki, N., Effects of L-carnitine on phospholipids in the ischemic myocardium, Jpn. Heart J. 28,243-251,1987.

6. Suzuki, Y., Kamikawa, T., Yamazaki, N., Effects of L-carnitine on ventricular arrythmias in dogs with acute myocardial ischemia and a supplement of excess free fatty acids, Jpn. Heart J. 45,552-559,1981.

7. Paulson, D.J., Schmidt, M.J.,Traxler, J.S.,Ramacci, M.A.,and Shug, A.L., Improvement of myocardial function in diabetic rats after treatment with L-carnitine, Metabolism 33,358-363,1984.

8. Pieper, G.M., and Murray, W.J., In vivo and in vitro intervention with L-carnitine prevents abnormal energy metabolism in isolated diabetic rat heart: chemical and phosphorus-31 NMR evidence, Biochem. Med. Metabol. Biol. 38,111-120,1987.

9. Pieper, G.M., Murray, W.J., Salhany, J.M., Wu, S.T., Eliot, R.S., Salient effects of L-carnitine on adenine-nucleotide loss and coenzyme A acylation in the diabetic heart perfused with excess palmitic acid. A phosphorus-31 NMR and chemical extraction study, Biochim. Biophys. Acta 803,241-249,1984.

10. Rodrigues, B., Xiang, H., and McNeill, J.H., Effect of L-carnitine treatment on lipid metabolism and cardiac performance in chronically diabetic rats. Diabetes 37,1358-1364,1989.

11. Molaparast-Saless, F., Liedtke, A.J., and Nellis, S.H., Effect of fatty acid blocking agents, oxfenicine and 4-bromocrotonic acid, on performance in aerobic and ischemic myocardium, J. Mol. Cell. Cardiol. 19,509-520, 1987.

12. Hekimian, G., and Feuvray, D., Reduction of ischemia-induced acylcarnitine accumulation by TDGA and its influence on lactate dehydrogenase release in diabetic rat hearts, Diabetes 35,906-910, 1985

13. Lopaschuk, G.D., Wall, S.R., Olley, P.M., and Davies, N.J., Etomoxir, a carnitine palmitoyltransferase 1 inhibitor, protects hearts from fatty acid-induced ischemic injury independent of changes in long chain acylcarnitine, Circ. Res. 63,1036-1043,1988.

14. Lopaschuk, G.D., Spafford, M.A., Davies, N.J., and Wall, S.R., Glucose and palmitate oxidation in isolated working rat hearts reperfused after a period of transient global ischemia, Circ. Res. 66,546-553,1990.

15. Paulson, D.J, Noonan, J.J., Ward, K.M., Stanley, H, Sherrat, A., and Shug, A.L., Effects of POCA on metabolism and function in the ischemic rat heart, Basic Res. Cardiol. 81,180-187,1986.

16. Lopaschuk, G.D., and Spafford, M.A., Response of isolated working hearts to fatty acids and carnitine palmitoyltransferase 1 inhibition during reduction   of coronary flow in acutely and chronically diabetic rats, Circ. Res. 65,378-387,1989.

17. Reinauer, H., Adrian, M., Rosen, P., and Schmitz, F.-J., Influence of carnitine acyltransferase inhibitors on the performance and metabolism of rat cardiac muscle. J. Clin. Chem. Clin. Biochem. 28,335-339,1990.

18. Shug, A.L., Thomsen, J.H., Folts, J.D., Bittar, N., Klein, M.I., Koke, J.R., and Huth, P.J., Changes in tissue levels of carnitine and other metabolites during myocardial ischemia, <u>Arch. Biochem. Biophys.</u> 187,25-33,1978.

19. Eistetter, K., and Wolf, H.P.O., Etomoxir. <u>Drugs of the Future</u>, 11,1034-1036, 1986.

20. Turnbull, D.M., Bartlett, K.,Younan, S.I.M., Sherratt, S.A., The effects of 2(5(4-chlorophenyl)pentyl)oxirane-2-carbonyl-CoA on mitochondrial oxidations, <u>Biochem Pharmacol</u> 33,475-481,1984.

21. Davies, N.J., McVeigh, J.J., Lopaschuk, G.D., Effects of TA-3090, a new calcium channel blocker, on myocardial substrate utilization in ischemic and nonischemic isolated working fatty acid-perfused rat hearts, <u>Circ. Res.</u> 68,807-817,1991.

22. Saddik, M., and Lopaschuk, G.D., Myocardial triglyceride turnover and contribution to energy substrate utilization in isolated working rat hearts, <u>J. Biol. Chem.</u> (in press).

23. Neely, J.R., and Morgan, H.E., Relationship between carbohydrate and lipid metabolism and the energy balance of heart muscle, <u>Annu. Rev. Physiol.</u>, 36,413-459, 1974.

24. Randle, P.J., Fuel selections in animals, <u>Biochem. Soc. Trans.</u> 14,799-806, 1986.

25. Patel, M.S., and Roche, T.E., Molecular biology and biochemistry of pyruvate dehydrogenase complexes, <u>FASEB J.</u> 4,3224-3233, 1990.

26. Kobyashi, K, and Neely, J.R., Control of maximum rates of glycolysis in rat cardiac muscle, <u>Circ. Res.</u> 44,166-175,1979.

27. Bremer, J., Carnitine - Metabolism and Functions, <u>Physiol. Rev.</u> 63,1421-1480, 1983.

28. Snoswell, A.M., and Koundakjian, P.P., Relationships between carnitine and coenzyme A esters in tissues of normal and alloxan-diabetic sheep, <u>Biochem. J.</u> 127,133-141, 1972.

29. Hansford, R.G., and Cohen, L., Relative importance of pyruvate dehydrogenase interconversion and feed-back inhibition in the effect of fatty acids on pyruvate oxidation by rat heart mitochondria, <u>Arch. Biochem. Biophys.</u> 191,65-81, 1978.

30. Lysiak, W.,Lilly,K., DiLisa, F., Toth, P.P., and Beiber., L.L., Quantification of the effect of L-carnitine on the levels of acid-soluble short-chain acyl-CoA and CoASH in rat heart and liver mitochondria, <u>J. Biol. Chem.</u> 263,1151-1156,1988.

31. McGarry, J.D., Robles-Valdes, C., Foster,D.W., Role of carnitine in hepatic ketogenesis, <u>Proc. Natl. Acad. Sci. USA</u> 72:4385-4388, 1975.

32. Vary, T.C., and Neely, J.R., Characterization of carnitine transport in isolated perfused adult rat hearts, <u>Am. J. Physiol.</u> 242,H585-H592, 1982.

33. Ichihara, K., and Neely, J.R., Recovery of ventricular function in reperfused ischemic rat hearts exposed to fatty acids, <u>Am. J. Physiol.</u> 249,H492-H497, 1985.

34. Liedtke, A.J., Demaison, L., Eggleston, A.M., Cohen, L.M. and Nellis, S.H., Changes in substrate metabolism and effects of excess fatty acids in the reperfused myocardium, <u>Circ. Res.</u>62,535-542,1988.

35. Lopaschuk, G.D. and Tsang, H., Metabolism of palmitate in isolated working hearts from spontaneously diabetic "BB" Wistar rats, <u>Circ. Res.</u> 61,853-858,1987.

36. Wall, S.R., and Lopaschuk, G.D., Glucose oxidation rates in fatty acid-perfused isolated working hearts from diabetic rats. <u>Biochim. Biophys. Acta</u> 1006, 97-103, 1989.

37. McVeigh, J.J., and Lopaschuk, G.D., Dichloroacetate stimulation of glucose oxidation improves recovery of ischemic rat hearts, <u>Am. J. Physiol.</u> 259,H1079-H1085, 1990.

38. Nicholls, T.A., Lopaschuk, G.D., and McNeill, J.H., The effects of free fatty acids and dichloroacetate on the isolated working diabetic rat heart. <u>Am. J. Physiol.</u> (in press).

DISCUSSION OF THE PAPER

D. LEHOTAY *(Canada)*: It has been postulated that free radical injury plays an important role in ischemia-reperfusion injury. Can the dichloroacetate induced increase in glucose oxidation prevent the injury completely, or is there an additional component that may be due to free radical injury that is not connected by dichloroacetate?

G. LOPASCHUK: It is clear that increasing glucose oxidation during reperfusion with dichloroacetate cannot completely protect the ischemic heart. If the ischemic episode is extended beyond 30 minutes, the depression of functional recovery in control hearts becomes even more severe. Under these conditions, dichloroacetate only partially improves functional recovery. It is clear, therefore, that other factors are contributing to ischemic injury. Whether the production of free radicals during reperfusion is one of the important factors mediating functional recovery has still not been completely resolved.

L. BIEBER *(East Lansing, MI)*: A comment - Micromolar amounts of etomoxir-CoA inhibit purified CPT-II, CAT & COT with 50% inhibition being between 1 to 10 μM. Your assumption of other inhibitions may be correct.

G. LOPASCHUK: Thank you for your comment! This would explain why in the intact heart, very high concentrations of Etomoxir no longer lowered cellular levels of long chain acylcarnitine.

F. DI LISA *(Italy)*: Can you establish a threshold for relating glucose oxidation and contractile recovery?

G. LOPASCHUK: We do not know exactly what level of glucose oxidation during reperfusion is necessary for the optimal recovery of function. As we demonstrated, addition of high levels of fatty acids decrease glucose oxidation from over 1500 μmol/g dry wt·min to approximately 150 μmol/g dry wt·min. We do know that although pharmacological doses of insulin increase glucose oxidation during reperfusion in the fatty acid perfused hearts by approximately 50%, this is not associated with an improvement in function. On the other hand, dichloroacetate, which has a protective effect, will triple glucose oxidation rates under these conditions. If I had to speculate I would suggest that optimal glucose oxidation rates during reperfusion are in the range of 300 to 600 μmol/g dry wt·min.

Conjunctive Enhancement of Adriamycin by Carnitine

A. Lee Carter, Randall Pierce, Carole Culbreath, and
Eugene Howard
Department of Biochemistry and Molecular Biology
Medical College of Georgia
Augusta, GA 30912-2100, U.S.A.

I. Introduction

Adriamycin (doxorubicin) distributed for clinical use in the United States by Adria Laboratories, Wilmington, Delaware is an antibiotic originally identified and isolated from a Streptomycete (1). Its structure and mode of action is similar to daunorubicin. Both are anthracycline glycosides and one of the best class of chemotherapeutic agents against a variety of cancers (2). The antineoplastic action of Adriamycin is thought to be caused by one or more of the following actions: a) Inhibition of RNA synthesis by template disordering and steric hindrance due to intercalation between stacked base pairs in the DNA double helix (3);
b) Direct inhibition of DNA polymerases by Adriamycin (4); c) Ultrastructural changes in nucleoli (5).

As an antineoplastic agent, Adriamycin has proven very useful in treating a number of different types of cancer which include leukemias, sarcomas, breast cancers and neuroblastomas (6-8). It seems to be a good drug to use in combination with other antineoplastic agents.

Severe side effects, in particular cardiotoxic effects, prevent more widespread use of Adriamycin for the treatment of cancer. The cardiotoxicity can be seen as abnormal electrocardiograms (9), histopathological changes (10) and often cardiac failure without any detectable symptoms (11). Early studies in cancer patients revealed severe cardiac problems often leading to death in 30-40% of patients receiving a total cumulative dose of Adriamycin which exceeded 550 mg/m^2 (12-13). Below this dosage, less cardiotoxicity was observed. Therefore, the total lifetime amount of Adriamycin given to patients has been restricted to 550 mg/m². If the cardiotoxicity could be prevented so that the dose could be increased above 550 mg/m², Adriamycin would probably be much more effective.

Using animal models, L-Carnitine administration has been successful in preventing cardiac damage or death from the cardiotoxicity of Adriamycin (14-18). These studies investigate the possible protective effect of carnitine towards the cytotoxicity of Adriamycin against human pancreatic cell lines. Also, the effects of Adriamycin and/or carnitine on the complexes I-IV of the mitochondial electron transport system in liver and heart.

II. Material and Methods

A. Materials: Adriamycin was a gift from Adria Laboratories, Columbia Ohio. Media and Media supplies were purchased from Gibco. Carnitine was a gift from Sigma Tau Pharmaceutical Co., Gaithersburg, MD. All other chemicals were of either reagent grade or the best grade available and were purchased from Fisher Chemical Co., Atlanta, GA or Sigma Chemical Co., St. Louis, MO.

B. Cell Cytoxicity Studies: Human malignant pancreatic cell lines derived by Dr. B. Chang (VAMC, Augusta, Georgia) from patients were used for these studies. These were of a low passage number and the cell lines have different growth characteristics. Cells were incubated at 37°C, 5% CO_2 in RMPI-1640 medium with 10% fetal calf serum and added glutamine. In the first experiment cells in log phase were incubated with different amounts of Adriamycin to determine the amount of Adriamycin required to kill 50% of the cells after 48 hours. This concentration of Adriamycin were used for subsequent experiments. Cells were also grown in the presence of different amounts of carnitine alone to determine whether carnitine affects the log growth rate of the cultures. Cells were then grown in the presence of LD50 Adriamycin and various amounts of carnitine for 48 hours to determine if cell survival increased in the presence of carnitine. Cells were trypsinized & live cells were determined by trypan blue exclusion.

C. Mitochondrial Electron Transport Studies: Sprague Dawley rats (100-125 g) were purchased from Harlan Industries, Indianapolis, IN. They were maintained in our facilities for three days before experiments began. Rats were housed in a 12:12 light:dark cycle and fed food and water *ad libitum*. They remained on regular rodent chow except for the carnitine deficient animals. Rats were maintained in a separate room for these experiments and were injected intraperitoneally as described for the various experiments. After a 2 to 5 day adjustment period the rats were randomly placed in treatment groups. The treatment groups were as follows: control, Adriamycin, carnitine, and carnitine plus Adriamycin. Animals receiving Adriamycin received 1 mg/k² body weight spread out over three days with 2 injections per day. Animals receiving carnitine received 100 mg/kg carnitine per injection on the same schedule as those receiving Adriamycin. All animals except those receiving Adriamycin plus carnitine received an injection of saline so that the volume injected per rat was equal.

Animals were sacrificed by carbon dioxide gas in the laboratory animal resources area using protocols approved by the MCG Commettee on Animal Use for Research and Education. The liver and heart were removed immediately and placed on ice.

Heart mitochondria from approximately ¼ of a rat heart were prepared as described (19). Briefly, the hearts were weighed and minced by a tissue grinder. Then the hearts were homogenized in a motor driven potter homogenizer with four volumes of 0.25 M sucrose. After homogenization, the sample was centrifuged at 1000 x g for 5 minutes. The pellet was resuspended in the sucrose solution and recentrifuged. The two supernatants were centrifuged at 12,000 x g for 15 minutes to pellet the mitochondria. If a large pellet containing contaminants was observed, the steps above were repeated. The mitochondria were dissolved in two ml of 0.05 M potassium phosphate pH 7.4 (KPi buffer) and sonicated to disrupt the mitochondria. This solution was used for the assay of electron transport chain complexes I-IV as described below.

Liver mitochondria were isolated as described (20-21). Briefly, livers from two rats were combined and homogenized with 8 volumes of 0.25 M sucrose solution. Mitochondria were separated from unbroken cells, cell debris and other organelles by differential centrifugation. Once mitochondria were isolated they were sonicated and prepared for assay as described above for heart mitochondria.

Assays of Complexes I-IV of the Electron Transport Chain: These assays have been described in detail (20-24). A brief synopsis of the procedures follows. Cytochrome C oxidase was measured by adding to 3 ml of KPi buffer an appropriate amount (usually 10 microliters) of mitochondria and establishing a baseline. Then, 10 microliters of cytochrome C solution was added and the rate of decrease in absorbance at 550/540 followed. NADH-cytochrome oxidase activities was monitored by measuring the reduction of cytochrome C at 550/540 following the addition of NADH to a solution containing KPi buffer, KCN, oxidized cytochrome C, and mitochondria. Succinate-cytochrome oxidase was monitored by measuring the reduction of cytochrome C at 550/540 following the addition of succinate to a solution containing KPi buffer, KCN, oxidized cytochrome C, and mitochondria. Succinate-Co-Q reductase was determined by measuring the reduction of DCPIP (dichlorophenol indophenol) at 600/700 following the addition of succinate to a solution containing KPi buffer, KCN, malonate, DCPIP, and mitochondria. NADH dehydrogenase was determined by measuring the reduction of potassium ferricyanide at 420/465 following the addition of NADH to a solution containing KPi buffer, potassium ferricyanide, antimycin A, KCN, and mitochondria. Succinate dehydrogenase was determined by measuring the reduction of potassium ferricyanide at 420/465 following the addition of succinate to a solution containing buffer, potassium ferricyanide, antimycin A, KCN, and mitochondria. Protein was determined by the method of Lowry (23) and all activities are expressed on a mg mitochondrial protein basis.

Carnitine (25) and Adriamycin (26) were assayed as previously described.

III Results

A. Effect of Carnitine on the Cytotoxicity of Adriamycin: By incubation of cultures of the individual cell lines with different amounts of Adriamycin (data not shown), the amount of Adriamycin required to kill half of the cells in a 48 hour period was determined. The concentrations of Adriamycin required for LD_{50} are as follows; PANI-20 nm/mi, PANII-30 nm/ml DX-65 nm/ml and SW - 100 nm/ml.

In the first experiment (Table I), carnitine was added at a concentration of 0.1 mM which is similar to the concentration found in pancreas, while in the second experiment (Table II) carnitine was added at 100 times the level in Table I. In this cell culture system carnitine did not protect cells against Adriamycin toxicity.

TABLE I

Cytotoxicity of Adriamycin with Physiological Concentrations of Carnitine

Cell type	Control	Adria	Carn	Adria $\pm$ Cam
PANC I	418$\pm$66	216$\pm$28	398$\pm$99	225$\pm$41
PANC II	319$\pm$86	137$\pm$29	325$\pm$43	135$\pm$54
DX	226$\pm$61	131$\pm$49	197$\pm$54	119$\pm$38
SW	112$\pm$31	38$\pm$22	123$\pm$18	58$\pm$17

The number of cells X 10^{-4} $\pm$ the standard deviation with 0.1 mM carnitine and 20-100 ng/ml of Adriamycin for 48 hours (see text above). Methods were as described above.

TABLE II

Cytotoxicity of Adriamycin with Physiological Concentrations of Adriamycin

Cell type	Control	Adria	Carn	Adria $\pm$ Cam
PANC I	512$\pm$102	267$\pm$92	553$\pm$158	261$\pm$60
PANC II	432$\pm$77	245$\pm$71	402$\pm$59	259$\pm$62
DX	214$\pm$61	115$\pm$40	246$\pm$46	123$\pm$43

The number of cells X 10^{-4} $\pm$ the standard deviation with 10 mM carnitine and 20-100 ng/ml Adriamycin for 48 hours (see text above).

B. Effect of Adriamycin and/or Carnitine on the Mitochondrial Electron Transport Chain: The dose of Adriamycin used in these experiments inhibited cytochrome C oxidase in heart but not in liver. No consistent effect was observed on the other activities of the electron transport chain in either heart or liver mitochondrial electron transport chain. L-Carnitine seems to protect cardiac cytochrome C oxidase from inhibition by Adriamycin.

TABLE III
Rat Mitochondrial Electron Transport Chain Activities
HEART

Activity	Control	Adria	Carn	Adria ± Carn
NADHDH X 10^{-2}	149±55	123±22	130±33	146±13
S-CytC R	478±159	505±154	403±269	555±282
NADH-CytC R	982±423	807±187	1300±662	1120±348
S-CoQ R	80±32	45±23	59±31	88±24
NADH-CoQ R	759±203	615±25	656±164	532±99
CytC O X 10^{-1}	214±110	79±14	129±97	144±56

LIVER

Activity	Control	Adria	Carn	Adria ± Carn
NADHDH X 10^{-1}	577±127	661±227	280±299	660±137
S-CytC R	63±30	168±72	115±75	124±25
NADH-CytC R	531±141	509±242	537±143	491±140
S-CoQ R	17±12	53±30	15±11	48±21
NADH-CoQ R	354±80	439±126	545±162	407±87
CytC O	188±135	339±193	105±25	165±66

The abbreviations are as follows: Adria, Adriamycin; carn, carnitine; NADHDH, NAD dehydrogenase; SDH, succinic dehydrogenase; S-CytC R, succinic cytochrome C reductase, NADHCytC R, NADH cytochrome C reductase; S-CoQ R, succinic coenzyme Q reductase; NADH-CoQ R, NADH coenzyme Q reductase; and CytC O cytochrome C oxidase. These are the means±standard deviation for 6-10 different rats. All assays were run in duplicate.

IV Discussion

This is the first report of the effects of L-Carnitine and Adriamycin on human cancer cell lines. Our data indicates that carnitine is not cytotoxic to pancreatic cancer cell lines and does not effect the cytotoxic properties of Adriamycin towards these cell lines. Subsequent experiments are needed to determine if L-Carnitine administration will affect the chemotherapeutic activity of Adriamycin In Vivo in tumor bearing animals or humans. Several reports indicate that L-Carnitine I.V. administration prevented the lethal effects of Adriamycin given in large doses (14-18).

While the exact mechanism of Adriamycin cardrotoxicity is not known, severe impairment of cytochrome C oxidase in heart could lead to sudden heart failure. These studies duplicate the impairment of cardiac cytochrome C oxidase by moderate doses of Adriamycin shown by others (27). However, hepatic cytochrome C oxidase was not inhibited by Adriamycin. No consistent effects were noted among the other activities of the mitochondrial electron transport system either in heart or liver. Concurrent administration of a typical dose of L-Carnitine partially prevents the inhibition of cardiac cytochrome C oxidase. This data is consistent with a recent report (28) that either D or L- carnitine prevents the depletion of a cardiolipin from cardiac mitochondrial membranes. The loss of cardiolipin results in reduced activity of Cytochrome C oxidase.

In summary, the data in this report support the hypothesis that the addition of carnitine to chemotherapeutic regimes that use Adriamycin may be useful in reducing the cardiotoxicity of Adriamycin while not reducing the cytotoxic effects of Adriamycin towards tumor cells. More studies are needed to determine the amount and frequency of L-Carnitine administration as well as human studies with different specific types of cancers to determine if the current maximum amount of Adriamycin could be exceeded.

V. REFERENCES

1. Chabner, B.A., Meyers, C.E., Coleman, C.N., and Jones, D.G. *New Eng. J. Med.* 292: 1159-1168, 1975.
2. Blum, R.H. and Carter, S.K. Ann. *Intern. Med.* 80:249-259, 1974.
3. Calendi, E., DiMarco, A., and Rediani, M. *Biochem. Biophys. Acta* 103:25-49, 1965.
4. Wang, J., Chervinsky, D.S., and Rosen, J. *Proc. Am. Assoc. Cancer Res.* 12:77, 1971.
5. Lambertenghi-Deliliers, G. In Carter, S.K., DiMarco, A., and Ghione, M. (eds.) *International Symposium on Adriamycin New York*: Springer-Verlag New York, Inc. 1972, pp. 26-34.
6. Wang, J.J., Chervinsky, D.S., and Rosen, J.M. *Cancer Res.* 32:511-515 1972.
7. Gottieb, J.A., Baker, L.H. and Quagliana, J.M. *Cancer* 30:1632-1638 1972.
8. Gottieb, J.A., Bonnet, J.D., and Hoogstraten. *Cancer Chemother. Rep.* 57:98, 1973.
9. Minow, R.A., Benjamin, R.S., and Gottlieb, J.A. *Cancer Chemother. Rep.* 6:195-201, 1975.

10. Billingham, M.E., Bristow, M.R., Mason, J.W., and Daniels, J.R. *Cancer Treat. Rep.* 62:865-872, 1978.
11. Ainger, L.E., Bushore, J., Johnson, W.W., and lto, J. *Natl. Med. Assoc.* 63:261-267, 1971.
12. Von Holt, D.D., Fayard, M.W., and Basa, P. Ann. *Intern. Med.* 91:710-717, 1979.
13. Lefrak, E.A., Pitha, J., Rosenheim, S., and Gottlieb, J.A. *Cancer* 32:302-314, 1973.
14. Vick, J., DeFelice, S., and Barranco, *Proc. Amer. Assoc. Caner Res.* 19:590-598, 1978.
15. Alberts, D.S., Peng, Y., Moon, T.E., and Bressler, R. *Biomed.* 29:265-268, 1978.
16. Breed, J., Ariaen, N., Zimmerman, J., Dormans, M., and Pinedo, H. *Cancer Res.* 40:2033-2038, 1980.
17. Van Fleet, J., Ferrans, V., and Weirich, W. *Am. J. Pathol.* 99:13-42, 1980.
18. McFalls, E.O., Paulson, D.J., Gilbert, E.F., and Shug, A.L. *Life Sciences* 38:497-505, 1986.
19. Tyler, D.D. and Gonze, J. *Meth. Enzymol.* X:75-77, 1967.
20. Alatefi, Y. and Rieske, J.S. *Meth. Enzymol.* 10:225-231, 1967.
21. Hogeboon, G.H. *Meth. Enzymol.* 1:16-19, 1955.
22. Hatefi, Y. *Am. Rev. Biochem.* 54:1015-1070, 1985.
23. Lowry, O.H., Rosenbough, N.J., and Farr, A.L. *J. Biol. Chem.* 193:265-275, 1951.
24. Dimauro, Bonilla, E., and Zeviani, M. *Ann. Neurol.* 17:521-528, 1985.
25. Maessen, P.A., Pinedo, H.M., Mross, K.B., and van der Vijgh, W.J.F. *J. Chromatog.* 424:103-110, 1988.
26. Carter, A.L., and Stratman, F.W. *FEBS Lett.* 111:112-114.
27. Praet, M., Calderon, P.B., Pollakis, G., Roberfro, M., and Ruysschan, I.M. *Biochem. Pharm.* 37:4617-4622, 1988.

ABSTRACTS TO PART IV

ISOLATION OF BUTYROBETAINE FROM BIOLOGICAL SAMPLES FOLLOWED BY DERIVATIZATION WITH 4'-BROMOPHENACYL TRIFLUOROMETHANESULFONATE AND HIGH-PERFORMANCE LIQUID CHROMATOGRAPHY. Paul E. Minkler, Stephan Krahenbuhl, and Charles L. Hoppel, Dept. of Veterans Affairs Medical Center, and Depts. of Pharmacology and Medicine, Case Western Reserve University School of Medicine, Cleveland, OH 44106.

4-(N,N,N-Trimethylammonio)-butanoate (butyrobetaine) was isolated from biological samples by extraction with methanol/acetonitrile, C_{18} solid phase extraction, and silica gel chromatography. The recovery of [^{3}H-*methyl*]butyrobetaine from spiked samples was 77% for liver, 76% for plasma, and 80% for urine. Samples were derivatized with 4'-bromophenacyl trifluoromethanesulfonate and the butyrobetaine 4'-bromophenacyl ester separated by HPLC. Radioactivity eluted in a single peak which co-chromatographed with standard butyrobetaine 4'-bromophenacyl ester (monitored at 254 nm) as a single, well resolved peak. The recovery of the injected radioactivity in this peak was 97.2% for liver, 99.1% for plasma, and 98.8% for urine. Two identical liver specimens were treated with this isolation procedure. Just prior to derivatization with 4'-bromophenacyl trifluoromethansulfonate, one was treated with butyrobetaine hydroxylase, while the second specimen was not. After derivatization, there was an absence of a butyrobetaine 4'-bromophenacyl ester peak and an increase in the carnitine 4'-bromophenacyl ester peak in sample exposed to butyrobetaine hydroxylase.

ACTIVITY OF BUTYROBETAINE HYDROXYLASE(BBOH) IN CHANNEL CATFISH. M.N. Carter, A.L. Carter, G. Burtle. Dept. of Biochemistry and Molecular Biology, Med. Coll. of Georgia, Augusta, Ga 30912 and Coastal Plains Exp. Sta., Tifton, GA.

The activity of BBOH was found to be at or below the limits of detection in assays of the livers of channel catfish fry, fingerlings, and mature catfish. Assays were also performed on pooled samples from six mature catfish to maximize activity measurement, however, there was no increase in activity.

A protein fraction containing BBOH was concentrated with mature catfish kidneys by ammonium sulfate precipitation. This experiment revealed a greater amount of activity than the liver experiments and 5% of the activity found in bovine tissue.

From these experiments we conclude that carnitine is synthesized to a lesser degree in channel catfish than in mammals.

CARNITINE THERAPY FOR CARDIOMYOPATHY.

Ken Jue MD, Susan Winter MD, *Linda Higashi and *Hugh Vance PhD.
Valley Children's Hospital and UCSF, Fresno, CA and *Metabolic
Research and Analysis, Fresno, CA.

Considering the roll of carnitine in fat metabolism it is not surprising that defects in carnitine metabolism result in cardiomyopathy (CMY). We recorded data on 33 patients who had CMY defined by echocardiogram and plasma levels of carnitine recorded prior to carnitine therapy. Thirty one patients had dilated and 2 hypertrophic CMY. Five had normal plasma free carnitine levels ($>20\,\mu$M/l), 7 had normal free levels with an elevated ratio of acyl to free carnitine (>0.40, insufficiency) and 21 were deficient ($<20\,\mu$M/l) with 14/21 also insufficient. In order to quantify the impact of carnitine therapy, we grouped the patients with regards to possible cause of carnitine deficiency/ insufficiency and scored them according to overall outcome. A score of 1, 2 or 3 indicates: no change, improved and resolved, respectively. Of 11 newborn (9 premature) the average score was 2.81 with 2 deaths. Of seven patients with diet deficiency, on parenteral nutrition or unsupplemented formula, the average score was 2.43. Seven patients with metabolic disorders had an outcome score of 1.86 with 4 deaths. Seven patients had no known cause for the carnitine deficiency/insufficiency and a score of 1.2.

INTERACTION OF CARNITINE WITH MITOCHONDRIAL CARDIOLIPIN.

V. Bobyleva, M. Bellei, E. Arrigoni Martelli', and U. Muscatello.
Institute of General Pathology, University of Modena, 41100, Italy, and 'Sigma Tau, 00040 Pomezia-Rome, Italy.

Studies by Mc Falls and coworkers have shown that doxorubicin causes a severe cardiopathy characterized by mitochondrial damages that can be largely prevented by administration of carnitine. Since doxorubicin is known to interact with cardiolipin, it was investigated whether the protective effect of carnitine may be related to an interference of the latter molecule at the level of doxorubicin-cardiolipin interaction. The results show that carnitine, both the L- and the D-form, activates mitochondrial respiration in the absence of added substrates, and that doxorubicin largely reduces, in the case of L-carnitine, or even abolishes, in the case of D-carnitine, this activation. This indicates that carnitine binds to the same site of mitochondrial membrane to which doxorubicin also binds and that this site is probably cardiolipin. If this is the case, the binding of carnitine to cardiolipin may induce a phase change with consequent modifications of the physical properties of the inner membrane.

DECREASED CARNITINE BIOSYNTHESIS IN RATS WITH SECONDARY BILARY CIR-
RHOSIS. <u>Charles L. Hoppel, Eric P. Brass, and Stephan Krahenbuhl</u> Dept. of Veterans Af-
fairs Medical Center, and Depts. of Pharmacology and Medicine, Case Western Reserve
University School of Medicine, Cleveland, OH 44106.

Carnitine homeostasis is altered in chronic liver disease. Carnitine biosynthesis
was investigated in rats with secondary bilary cirrhosis induced by bile duct ligation (BDL)
and pair fed sham operated rats (CON) fed an amino acid-based diet. The body weight
and total body carnitine content were similar, but the urinary excretion of carnitine and
butyrobetaine (BB) were decreased by 50% in BDL as compared to CON.

	Carnitine	Butyrobetaine	Trimethyllysine
	Net Balance (umoles / 100 g body weight / day) Mean +/- SD		
CON	0.93 +/- 0.08	0.99 +/- 0.09	1.60 +/- 0.12
BDL	0.45 +/- 0.19	0.48 +/- 0.19	1.91 +/- 0.40

Total trimethyllysine (TML) production was unaffected by BDL. However the urinary excretion of TML was increased with corresponding decreases in production of BB and carnitine. BDL rats thus demonstrated a defect in carnitine biosynthesis at the level of conversion of TML to BB. Total body carnitine was preserved with BDL secondary to renal carnitine conservation.

EFFECT OF CARNITINE ON CALCIUM DEPENDENT UNCOUPLING OF
LIVER MITOCHONDRIA INDUCED BY CARBOXYATRACTYLOSIDE AND
FATTY ACIDS E. Arrigoni-Martelli[*], E. Mokhova [t], M.
Bellei, U. Muscatello and V. Bobyleva Institute of
General Pathology, Modena Italy, *Sigma-Tau, Pomezia,
Rome, Italy [t]A.N. Belozersky Laboratory of Moscow State
University, USSR.

The effect of fatty acids and of
Carboxyatractyloside on respiration of Ca^{2+} -loaded
mitochondria and the protective effect of carnitine have
been studied. The omission of ATP from incubation medium
abolished or strongly decreased the ameliorating effect
L-carnitine, suggesting that in our experimental
conditions carnitine mainly protects against damaging
effect of fatty acids. In freshly isolated (but not in
the aged mitochondria) 5mM D-carnitine also showed a
protective effect on the retention of calcium by
mitochondria. 5 µM oleate, 10 nM FCCP and
Carboxyatractiloside (at concentrations at which it
specifically interacts with antiporter) decreased the
Ca^{2+} retention to about the same degree, indicating that
protonophoric effect of fatty acid may be essential for
Ca^{2+}-dependent initiation of mitochondrial damage.

Index